T0176585

FILAMENTARY ION FLOW

FILAMENTARY ION FLOW
Theory and Experiments

FRANCESCO LATTARULO

VITANTONIO AMORUSO

IEEE PRESS

WILEY

Library of Congress Cataloging-in-Publication Data:

Lattarulo, Francesco.
    Filamentary ion flow : theory and experiments / Francesco Lattarulo,
Vitantonio Amoruso.
        pages cm
    ISBN 978-1-118-16812-7 (cloth)
  1. Ion flow dynamics. 2. Electrostatics.   I. Amoruso, Vitantonio, 1955- II. Title.
    QC717.L38 2013
    537'.2–dc23
                                                                    2013029213

Printed in the United States of America

10 9 8 7 6 5 4 3 2 1

There is nothing more uncommon than common sense.

Anonymous author

Because, up there, in heaven, isn't paradise an immense library?

Gaston Bachelard (La Poétique de la rêverie)

CONTENTS

PREFACE

The arguments put forward by this book offer a coupled theoretical framework for an appropriate description and treatment of unipolar ion flows subject to electric fields. Several mechanisms can be adopted to generate ion flows, but exclusive reference is made here to those typical charge-injecting sources that are confined to overstressed (in an electrostatic sense) surface areas of conductors under direct current corona. The rationale for this choice is that a pair of main requirements for the physical phenomenon to be investigated, namely, generating ions and then setting them in motion by an impressed force, are met at once. The ion source identifies with an ionization region that occupies a restricted volume of space in contact with a conductor raised at a potential exceeding the corona onset level. Therefore, it could be said that the far broader drift region, where the repelled ions can slowly flow toward a collecting counterelectrode, covers the entire electrode gap. Under the described circumstances, and with special reference to the ion flow crossing the drift region, the governing electromagnetic and fluid dynamic laws form that special body of knowledge preferentially referred to as electrohydrodynamics (EHD). Given that there is a large amount of traditional and emerging practical applications somehow involving EHD applied to drifting ions (see later on), urgent improvements to available theoretical resources need to be made. Before going straight to the heart of the matter, it is positive to say a few preliminary words on the difficult task of finding a way of moving forward in the world of interdisciplinary investigations, where the present one is a prominent example. Any skillful and prudent researcher is aware of the fact that carefully putting together a set of governing equations under the described complex situations could result in a very hard undertaking, sometimes exposed to misinterpretations and inconsistencies. The excuse for this difficulty resides in the need of getting, at first, information from broken up sections of physics before attaining a final joint model. It has been argued at times that a physical problem appears to be far too involved not owing to the true attributes of the overall system, but to the mathematical aspects of the defective model being profiled. Note further that a good grounding in reasoning on composite problems could in principle allow the emergence of some still submerged relationships between parameters that are traditionally pertaining to different areas of physics. Unfortunately, that is not really the case because when the departure between theory and experiment becomes a matter of some importance, the improving efforts often adopted by trial and error consist in supplementing the starting model with a number of higher-order terms. Such a commendable intention could result in a frustrating, unfruitful exercise because first-order interactions between parameters distinct from and usually handled in different compartments of physics still remain overlooked. These disappointing circumstances seem to characterize previous EHD approaches because the electromagnetic and fluid dynamic laws, somehow involved in the description of unipolar ion flows, are kept substantially decoupled. Very specifically, this is the case for the electric and velocity fields, enlivened in the ion-drift region, which are usually assumed to respectively exhibit divergence and circulation both different from zero. The real value of the present treatment essentially consists in highlighting the deficiencies commonly compromising EHD models and opposing them with a reformulated modeling that carefully takes into account mutual influences between the involved field laws. This revised approach distinctively applies in the drift regions because of the subsonic velocity of the ions flowing there. Indeed, all it requires is to only overcome some conceptual drawbacks resisting the explored hybridization. This will ultimately be expressed by the solenoidal and irrotational vector properties that, of necessity, the respective electric and velocity fields simultaneously should gain. Surprisingly and paradoxically, there is an added value to the given coupling: The complete governing formulation looks quite unsophisticated and eloquent to the benefit of physical interpretation. Especially in view of engineering applications, a recommended finishing touch is that of cautiously using Occam's razor to cut out the often useless, confounding, and oversophisticated higher-order terms mentioned above. Consider that long-lasting laboratory tests and mature reflection on the supplied databases ultimately persuaded this investigator on the validity of the claimed hybrid model. The adopted theoretical scheme has been derived imparting, after R. E. Kalman, in that lucid advice that reads "get the physics right, the rest is mathematics,"[*] which is thus rather against the largely complied with and competitive advice that could read "get the mathematics right, the rest is physics." The compelling attraction exerted by the latter approach is, in general, questionable since the adopted mathematical structures invariably work well only in the confined domains of physics where they were conceived. Instead, a correct mathematical approach to an integral reality needs to be founded on a common theoretical substrate whose specialized aspects are permissible to the extent that it holds unbroken.

Perhaps the more striking aspect of the present coupled treatment is that a subsonic ion drift is instead seen to assume the discontinued configuration of a filamentary flow guided by a Laplacian-field pattern. This issue is in contrast with previous uncoupled theories according to which a space-filling ion flow is claimed to cross the drift region under the action of a Poissonian electric field. The raised difference is, in this author's opinion, the very cause of the commonly perceived departure between observables and theoretical predictions. This drawback is becoming increasingly unacceptable in view of the latest breakthroughs into avionics and turbomachinery, in relation to levitation or air propulsion by plasma actuators, and into geophysics and electromagnetic compatibility, in relation to ionosphere and pre-stroke mechanisms, to name just a few. Parenthetically, everything still remains to be done in evaluating whether this multichanneled corona-driven flow can be either fruitfully exploited in microfluidics—that is to say, as a substitute for diffusion-driven flow in microfluidic channels—or play a role in those geoelectric phenomena, classifiable as earthquake by-products, interfering with grounded sensitive systems. Even more traditional applications related to charge transfer—thus involving HVDC line environment, electrostatic precipitators, lightning protection systems, electrophotography, dry powder coating, ESD, ignition hazard, surface treatments of materials (fabric, etc.), and ozonizers—must be able to reap the benefits of research that has been carried out and reported in this book.

Several years ago, taking advantage of the sympathetic disposition of my collaborator Vitantonio Amoruso, I decided to bring together our experimental efforts to carefully understand the phenomenological aspects of the raised difficult problem. To this end, a patrimony of data, resulting from a three-decade activity developed in an *ad hoc* arranged department high-voltage laboratory, has organically been supplied. This has been made feasible by the special use of some unusual electrode assemblies. As a result, unexpected, submerged, or obliterated phenomena have been discovered and carefully taken into account in order for this unprecedented theory to be substantiated

---

[*]Opening lecture IFAC World Congress—Prague, July 4, 2005.

by convincing arguments. Therefore, the content of the book is preferentially addressed to all those who are employed in the forefront of research or are always willing to question currently available design paradigms.

The monograph is subdivided into five chapters, the first two and the remaining three separately authored, respectively, by my colleague Vitantonio Amoruso and the undersigned. Chapters 1 and 2 are written in the form of cursory commentaries on the basic phenomenology of corona activity (Chapter 1) and ion-drift theories (Chapter 2), whereas underlying principles of fluid dynamics are found in Chapter 3. The arguments treated in Chapters 1–3 are preliminary to the pair of key Chapters 4 and 5, the former providing deep insights into the physics of ion flows crossing drift regions by a joined theory, and the latter giving the necessary experimental support. It is suggested that the reader should take a look at the Introduction, where a more extended presentation of the chapter content is furnished in two respects: The reader will be able to better understand the scope and rationale of the book and comfortably discern those issues that might be of most interest to him or her.

FRANCESCO LATTARULO

ACKNOWLEDGMENTS

On behalf of my coauthor, I would like to thank all the students who worked on their graduation theses on this subject and collected essential databases for the figures of Chapter 5 to be traced. In particular, I am grateful to F. V. Ambrico, R. Bronzino, G. Bulzis, D. De Luca, C. Florio, M. Rubini, and P. C. Schiavone for providing Figures 5.21; 5.27; 5.22–5.25; 5.6 and 5.7a; 5.17–5.19; 5.7b, 5.11, 5.14; and 5.10 and 5.13, respectively. Often reference is made to his/her life partner for help in checking the written material. This is not the case for my wife Gabri, merely because our common mother tongue is not English. Notwithstanding this, I have the good fortune of experiencing how she successfully contributes to achieving any objective in a mysterious way. Last, I owe a great debt of gratitude to Mary Hatcher, Associate Editor at John Wiley & Sons, for having commendably committed herself to the preparation of this book.

F. L.

INTRODUCTION

This book is divided into five chapters, (four of) which include one or more Appendices. The complexity of the subject matter accounts for the number of sections in each chapter. It is important to clarify that the purpose of this Introduction is not merely to summarize the separate items in an orderly fashion, but to provide an appropriate framework to enable the student to learn. Therefore, directly following a description of the content, the reader will find a brief comment about the reasoning behind the order of the material presented. The authors are confident that creating logical links among individual arguments in different sections of the book is the best way to present them as a whole. It is hoped that this will lead to improvements in technological applications traditionally accommodated in the wide realm of applied electrostatics, namely, wherever the notion of charge transfer applies to some extent. But that does not alter the fact that the book is especially tailored to the needs of individuals who are prone to examine the past with a critical eye and to create innovative ideas. Therefore, at the end of this Introduction, some suggestions are made that look far enough ahead at recent or unfamiliar applications where the theory can be appreciated.

Chapter 1 provides the reader with essential information about gas discharges, with special attention paid to different corona modes. These manifest when the designated active conductor is raised at a stationary potential with assigned magnitude and polarity. Several kinds of ionization processes and streamer mechanisms are addressed. The important role that photoelectrons and photoionization of gas molecules assume surrounding avalanches for self-propagating streamers to develop is stressed. Also, a description of

Kaptzov's hypothesis applied to the electric field on the surface of active conductors is presented.

Chapter 2 is devoted to theoretical results, often achieved by experimental tests, derived from a collection of investigations on corona-originated unipolar ion flows and related effects. Usually, the Poissonian electric-field strength, charge density, and current density are taken as individual or collective surrogates in describing the electrical properties of ionized flows. The investigation is performed inside the drift region and, more often, on the inactive electrode because this has to play the dual role of collector for the impacting ions and measuring surface for the three parameters mentioned above. It is stressed that dealing with computational electrostatics of steady ion flows would imply resolving a homogeneous, fully nonlinear third-order partial differential equation. Apart from the difficulty in appropriately assigning all the boundary conditions to which the calculation is subject, the solution is an impossible task. This explains why it has not been concluded, because this survey makes it quite clear that the already long list of those that were and even currently are involved in attempting computational strategies—all invariably debatable for some reason—has been aimed at circumventing insuperable theoretical difficulties. As will be shown, use is basically made of the charge-drift formula and/or simplifying hypotheses applied to the boundary (notably, Kaptzov's hypothesis already introduced at the end of Chapter 1), to the ion trajectory pattern (this is the case for the familiar Deutsch's hypothesis conveniently introduced here), to the physical properties of the gaseous medium (extensively discussed in Chapters 3 and 4), and so on. Even more advanced computational resources, devised to do something about the far more difficult problem of waiving the above facilities, are often represented by arbitrary derivation of some boundary conditions.

Chapter 3 describes the basic notions of fluid dynamics that qualify as functional to the joined model that is being formulated. Conservation laws for mass, momentum, and energy, along with related ones, are conveniently expressed in differential form. The key notion of subsonic flow is introduced and, hence, due account is taken of enforcing the velocity field of the fluid particle to become solenoidal (a source-free one). This condition is then accomplished, for the reasons that will be explained in the next chapter, while also imposing a curl-less character on the flow. Combining the above pair of properties for the flow velocity is proved to be consistent with ionized particles slowly moving along preestablished Laplacian-field streamlines toward the collector. A reading of such a specialized dynamics is given, and identification of the constant energy density of the bulk motion with the kinetic energy is stressed. This issue is fully consistent with ionized particles flowing with zero conservative body forces, namely, as a result of some momentum transfer, by molecular collision, to the springing up ions before being injected into the field domain. It will be clear that in the end the above circumstances are equivalently interpreted as giving the gaseous fluid that fills the drift space excitation-field-independent properties, an issue of paramount importance exactly in view of the present application.

Chapter 4 describes an original coupled model for corona-originated unicharge flows. Starting from the primal, combinative notion of reduced mass-charge applied to the totality of gas particles forming the fluid, the common justification for classifying conduction in gases (this is the case for ion drift) and convection of charged molecules into two groups is then removed. As an overwhelming consequence, a kind of mutual influence between ion-drift velocity $\mathbf{v}$ and electric field $\mathbf{E}$ may occur, and that would explain their reciprocal identification. This is permitted to within a trivial factor, standing for ion mobility k, since a widely adopted low-field approximation pertaining to this investigation allows the physical relationship $\mathbf{v} = k\mathbf{E}$ to hold. As extensively illustrated, the filamentary structure so given for the ionized flow is the logical consequence of the combined solenoidal and irrotational nature of the individual $\mathbf{v}$- and $\mathbf{E}$-fields that are about to be joined. In brief, any elemental current-carrying channel of the given multichanneled flow model behaves as a filamentary conductor along which the electrodynamic $\mathbf{E}$-field and the ion velocity $\mathbf{v}$-field are unidimensional and uniform. This appealing picture clashes with previous theories on ion drift and related arguments, even though some commendable attempts are also made here to regain them somehow. The raised disagreement is entirely expressed through those remarkable differences of a morphological nature that this coupled flow model shows in comparison to counterparts available to a large extent, all of them suffering from being uncoupled. In fact, the ion flow is enforced in the former model to become discontinued (in a filamentary sense) while being guided by a Laplacian-field pattern, whereas a continuous (in a space-filling sense) Poissonian-field pattern invariably derives from the latter. The treatment points out how uncoupled models have come under criticism since, even as they do not query the validity of the relationship $\mathbf{v} = k\mathbf{E}$, they actually contradict this equality. In fact, admission is made in uncoupled models for $\mathbf{E} = \mathbf{v}/k$ to be a nonzero-divergence field, an attribute inconsistent with the subsonic classification of ion drifts invariably subject to solenoidal $\mathbf{v}$-fields. In particular, this chapter offers opportunities for both (a) wide-ranging discussions on the very good reasons to finally substantiate Deutsch's hypothesis (DH) and (b) properly addressing the question of ion wind's source and blowing.

Chapter 5 gives substantiation to the cardinal equation [Eq. (4.25)] on which the filamentary ion-flow theory is founded. The experimental approach has been made possible by using some uncommon electrode assemblies and adopting *ad hoc* performed remote monitoring for suitable databases to be furnished. After these have been collected and carefully examined, a connection between the ion injecting mechanism and overall structure of the Laplacian field in the drift region is proved. As pointed out, even with the assistance of invariance principles, this discovery has been the deciding factor in safely claiming a filamentous nature for the category of ion flows.

Each chapter, with the exception of Chapter 1, is supplemented with Appendices treating the specific subjects summarized as follows:

- Appendix 2.A—Warburg's cosine law customarily applied to the rod-plane corona.
- Appendix 2.B—Bipolar ionized field, for the sake of completeness (bear in mind that monopolar ion flow is the main subject of the book).
- Appendix 3.A—Thermodynamic quantities irrespective of the gaseous medium's (namely, at rest or subsonically moving) status.
- Appendix 4.A—Emitter–collector relationships, involving ionized field parameters, through the key definition of channel density applied to the multi channeled flow model at hand.
- Appendix 4.B—Governing equations for diffusional effects and related influence on the boundary conditions.
- Appendix 4.C—Viscosity–mobility relationship for gases, after Walden's rule applied to liquids, an issue consistent with the coupled model.
- Appendix 4.D—Substantial derivatives involved to theoretically show inherent consistency of the relationship $\mathbf{v} = k\mathbf{E}$ with the expected $\mathbf{v}$- and $\mathbf{E}$-fields' potential distributions.
- Appendix 5.A—Subsidiary theoretical arguments to sustain generalization of Warburg's law (see also before Section 2.4.1 and Appendix 2.A). Accordingly, exponent n of the common cosine law is proved to assume specific integers, at least for the 2D cases studies listed in Table 2.1.

For the reader's convenience, some links are made according to the following outline:

- Owing to the interdisciplinary character of the treated subject, those who are familiar with electromagnetism but have no knowledge of fluid dynamics may be surprised to find some terminological differences in describing common vector properties. This is especially the case in reading Chapter 3, where use of the locution "vorticity" of an unspecified vector $\mathbf{F}$ in the substitution of "circulation" of $\mathbf{F}$, both to mean rot $\mathbf{F}$. Additionally, an isochoric flow is nothing but a solenoidal (or source-free) flow in the $\mathbf{v}$-field, for which div $\mathbf{v} = 0$ ($\mathbf{v}$ stands for flow velocity).
- The set of potentials applied, in particular, to the partially covered wire (Section 5.5) allow photoionization-promoted initiatory phenomena

(Section 1.2.4) occurring in some receded locations external to and all around the cover to be neglected. This permission cannot be denied against a comparison with that strong activity concentrated on the uncovered fractional amount of the entire wire surface corresponding to the slot. Such an observation is made as an attempt to focus, by elimination, on a tangential spreadout of ions from the slot, a phenomenon supposedly dominated by an initially strong diffusional thrust prior to their outward injection (Sections 4.3.1, 4.4.2, and 4.6 and Appendix 4.B).

- In spite of a pulsative corona activity (Sections 1.9.2 and 1.9.3), the elementary pulses may be assumed to merge in time (Appendix 4.B), compatibly with the applied voltage. Provided that this remains unchanged, the ion flow appears substantially steady, as the corona current does even at lower voltages (here and there in Sections 4.3, 4.4.2, and 4.5.2). This implies that time-independent governing equations for the potential of the electric field in the presence of space charge (Sections 2.2.1 and 4.3), and related parameters, are adopted. Even admitting that unsteady dynamics can occur, the contribution of the variable magnetic field is expected to be comparatively unimportant for the present class of problems (Section 4.3.5).

- The otherwise tested and widely adopted low-field assumption (Sections 4.2 and 4.3.7) for the equality $\mathbf{v} = k\mathbf{E}$, with k denoting a constant mobility (Section 2.1), tacitly gives arguments to substantiate the coupled model subject to rot $\mathbf{E}$ and div $\mathbf{E}$ both dropping to zero (Section 4.5.2 and Appendix 4.D). It is common knowledge that the first condition is distinctive of electric fields, whereas it is quite disregarded elsewhere that even the second condition legitimately applies. This by virtue of the novel definition of reduced mass-charge (Sections 4.1–4.3), which leads to div $\mathbf{E} = 0$ for the electric field in an ion-drift region. Here, the velocity is notoriously subsonic, whereby div $\mathbf{v} = 0$ and, in turn, div $\mathbf{E} = 0$ simultaneously apply (Sections 3.2, 3.9.2, 3.10, 3.11.3, 3.12.1, and 3.12.2 and Appendix 3.A).

- As previously mentioned, ion flows rigidly guided by Laplacian fluxlines (this is a prominent feature in the present treatment) are in fact fully responsive to DH, an argument around which ion-flow models procrastinated for far too long (Sections 2.3 and 2.6). An in-depth analysis now enables us (Section 4.3.8) to pursue the penultimate aim of incidentally proving its definitive validity and the ultimate aim of giving further arguments in favor of the filamentary model. Overall, this object has been achieved in an articulate manner, first remembering that real gases under ordinary ambient conditions can be safely treated as being ideal and then verifying that the thermodynamic quantities associated with the flow hold constant in practice (Appendix 3.A). As can be seen, the inquiry into the cause of DH has to progress through a series of arguments involving fluid dynamics as well (additionally, see Sections 3.5.3 and 3.7). This turns out to be the prerequisite for a self-consistent final formulation of the coupled ion-flow model.

- Those who are aware of the difficulties that arise in attempting direct detection of the injection mechanism and consequent low-energy ion-swarm scenario know full well the value of indirect laboratory tests. These have been profitably performed to ultimately sustain the hard inverse problem of reconstructing distribution laws at the emitter (Appendix 4.A) by distanced measurements at the collector (Sections 5.2–5.7).

As already mentioned in the Preface, prominent importance could be currently ascribed to prototyping some devices for avionic applications. The spotlight is now on the electrostatic levitator and plasma actuator—the former presumably, the latter surely, based on the ionic wind for air propulsion. In addition, usually performed performance surrogates, as with pressure ratio, efficiency, and rotor power for some rotating engine studies, are also theoretically evaluated in understanding how the flow control actuation can be exploited in the best possible way. A reference is reported here, in the presentation of Chapter 4, toward the end, regarding the byproduct of corona activity represented by the ionic wind. Briefly, we have to look at the whole question that there is a need to differentiate the motion of the charge carriers under examination, whether it is subsonic or sonic/supersonic. With particular reference to levitation, it is a key point that should give rise to a more thorough study of the ion wind's genesis and blowing. Putting it bluntly, this is about being aware of the fact that as soon as the supersonic ion velocity switches to subsonic, namely, when recently formed (in the ionization region) unicharges with the same polarity of the active conductor are injected into the wider drift region, then the ion wind is predicted here to be generated only at the interface between those regions. This as a result of a type of momentum-transfer mechanism concentrated in close proximity to the active electrode because of the short dimensions of the ionization region. Therefore, such a description conflicts with previous expectations according to which levitation forces are instead permitted to enliven ion wind inside the drift region, throughout. As an immediate consequence, even cutting-edge 3D packages implemented for a theoretical simulation of device attributes cannot be immune to criticism and, hence, cannot prevent concerted design paradigms from being revisited. Currently, this is a cause for concern since the above computational codes show limitations in capturing the effects under examination. As regards the efficiency of a type of plasma actuator, the abrupt passage from the ionization to drift region could be separated by the ionic wind flow by constructional artifices. But the fact remains that carefully describing the morphology of the ionization region can importantly contribute to efficiency estimations even when the actuation is essentially due to an alternative electric field causing a dielectric-barrier discharge plasma. To this end, the subject matter treated in this book can be complemented with the electron component for the active zone to be theoretically analyzed. In any case, the model gets insights into the strong dependence of the plasma performances on some key factors for which the designer must have due regard. For instance, reference could be made to the role that diffusion can play in steadily shaping the peripheral region of the plasma. Additionally, it may be that some unwanted subsonic flows of charges, embedded somewhere in that region presumably occupied above all else by supersonic flows, can as such significantly modify the overall performances of the prototype. Passing now to a different problem, consider that electromagnetic interferences involving sensitive components could also be those caused by geophysical phenomena in the atmosphere, even those occurring at the Earth's level. The victims are on-board and on-ground electric power and communication systems that happen to be installed where a thunderstorm or tectonic activity is taking place, or thereabouts. The most important effect connected to the former activity is lightning and related preliminary phenomena. In particular, appropriately addressing pre-stroke models is a crucial exercise finalized to estimate the expected preventing or intercepting features of air terminals usually adopted as lightning protection systems. Of course, the same models can be used to predict preliminary mechanisms involving inadvertent interceptors. Interpreting the active head of the descending leader as the endpoint of a current impressed slanting filament could be useful in better investigating the intercepting attributes of intentional and unintentional structures. Therefore, the investigation cannot be performed without relying upon a suitable ion-drift model. With special interest in the second case study, which is an unexplored one, let the upper half space be exceedingly ionized by radon efflux presumably accompanying earthquake episodes, and let a background electric field be simultaneously present. Under these circumstances, ion flows formed in the atmosphere can induce an Earth-surface potential. Spurious currents are permitted to carry down the length of lines where mutually distanced and grounded susceptible equipments can be connected. Even though the extent and severity of this interfering mechanism vary considerably, the subject may be of interest for electromagnetic compatibility investigations. Because geoelectric phenomenology is quite similar, in a qualitative sense, to that determined by exceptional geomagnetic storms, it deserves attention.

PRINCIPAL SYMBOLS

a	electrode spacing; emitter elevation (unless otherwise specified)
a_p	characteristic length of a fluid particle
A	action
A_0	neutral gas atom in the normal state
A_T	Townsend formula constant
A^*	atom in the excited state
A^{**}	metastable
b	inter-axis distance between parallel cylinders
b	coordinate
$\mathbf{B}$	magnetic flux density
B_0	neutral gas atom in the normal state
B_T	Townsend formula constant
c	speed of light in free space
c_m	dimensionless coefficient
c_s	local speed of sound
d	diameter
d	cathode-to-anode distance
d_x	coordinate
dA	streamtube or fluxtube cross-sectional area
dA_m	molecule surface area
dV_m	molecule volume
D	diffusion coefficient
$\mathbf{D}$	rate-of-strain tensor; electric displacement

$\frac{D}{Dt}$	substantial (or material) derivative
e_i	specific internal energy
e_k	kinetic energy per unit mass of the bulk motion
e_t	total kinetic energy per unit mass
E	electric field strength
$\mathbf{E}$	electric field
$\mathbf{E}_L$	Laplacian electric field
E_0	detected corona onset electric field
$\mathbf{f}_m$	body force
$\mathbf{F}$	force
h	sphere center suspension; emitter elevation (in some cases, instead of a)
h	Planck's constant
h_f	focal length of confocal ellipsoids
h_0	height of a spherical equipotential with short radius
h_s	height of the center of a spherical electrode
$h\nu$	photon quantum of energy
$\mathbf{H}$	magnetic field
i_c	steady current carrying a thin wire
I	channel carrying electric current; corona current
$\mathbf{I}$	unit tensor
IP	injection point
$\mathbf{J}$	current density
K_B	Boltzmann's constant
k	ion mobility
K	dimensionless constant
K'	dimensionless constant
K_n	Knudsen's number
l	dimensionless scalar function of position
ℓ	scalar function of position
L	fluxtube median line length; streamline, fluxline, or ion trajectory length
L	Lagrange function
L_D	diffusion distance
L-	longitudinal-type
m	roughness factor
m_c	ion mass
m_0	mass of a single molecule
M	Mach number; mass of neutrals
M_e	number of secondary electrons released at the cathode per emitted primary electron
n	number density

$\mathbf{n}$	outwardly directed unit vector
n_c	number of electrons emitted from the cathode surface; ion number density
n_0	initial number of electrons emitted from the cathode
n_0'	number of electrons emitted by secondary ionization processes
n_x	increased number of electrons at distance x from the cathode
N	total number of channels
N_c	total number of primary ionizing collisions in the gas per primary electron emitted from the cathode; critical size of an avalanche
p	thermodynamic pressure
P_e	Péclet's number
q_c	ion's charge
$\mathbf{q}_T$	heat flux density vector
r	radius
$\mathbf{r}$	particle position
r_0	short radius of a spherical equipotential
r_s	radius of a spherical electrode
R	radius
R_i	ionic recombination coefficient
RSP	referential source point
S	surface
$\mathbf{S}$	tensor
t	time
T	fluid absolute temperature
$\mathbf{T}$	Cauchy stress tensor
T_k	total kinetic energy
T-	transversal-type
$\mathbf{u}$	force unit vector
U	constant energy density of the bulk motion
$\mathbf{v}$	velocity (modulus v)
$\mathbf{v}_D$	diffusion velocity
$\mathbf{v}_m$	gas velocity
$\mathbf{vv}$	dyadic
V	volume
V_b	breakdown voltage
V_0	potential of a spherical equipotential with short radius
V_T	corona onset voltage
w	temperature-dependent average speed of a molecule
$\boldsymbol{w}$	wind speed
w_e	atom excitation energy
w_i	atom ionization energy

w_k	average kinetic energy acquired by a particle carrying a charge q
W	potential energy per unit volume
W_E	electrostatic potential energy density; work of charging
W_t	total positional energy density
W^{**}	energy of a metastable
x	coordinate
x_0	coordinate
x_ξ	coordinate
y	coordinate
α	isothermal compressibility coefficient
α	primary ionization coefficient
α_e	number of ionizing collision per unit length made by an electron
α_i	injection angle
α_j	cone angle
α_t	tilting angle
α_v	inclination angle
$\bar{\alpha}$	effective ionization coefficient
β	bulk expansion coefficient; azimuthal angle
γ	angular deviation
γ	efficiency of secondary electrons' emission process (Townsend secondary coefficient)
Γ	streamline, fluxline, or ion trajectory length
Γ	generic tensor
$\delta\mathbf{l}$	oriented segment of a linear particle
ε	medium permittivity
ζ	collision period; dimensionless factor labeling a circular fieldline
η	attachment coefficient
θ	semi-vertical cone or wedge angle of discharge
λ	mean free path
λ_p	radiated photon wavelength
μ	viscosity; medium permeability
μ'	second coefficient of viscosity
ν	kinematic viscosity
ξ	angular deviation
ξ_a	dimensionless factor
ρ	reduced-charge density
ρ_c	ion charge density
ρ_m	mass density; reduced-mass density
σ_T	fluid heat conductivity

τ	tangential shear stress; time-of-flight
τ_D	diffusion time
$\boldsymbol{\tau}$	external force per unit surface
υ	proportionality factor
$\boldsymbol{\Phi}$	unspecified agent
$\boldsymbol{\Psi}$	unspecified agent
φ_E	Poissonian electrostatic potential
φ_L	Laplacian electrostatic potential
$\varphi_v(\mathbf{r}, t)$	Stokes' scalar potential for the velocity
$\boldsymbol{\psi}_v(\mathbf{r}, t)$	Stokes' vector potential for the velocity
Ω_0	ion's collision cross section
ω	equipotential line parameter
∇^2	Laplace operator

CHAPTER 1

FUNDAMENTALS OF ELECTRICAL DISCHARGES

1.1 INTRODUCTION

Natural phenomena, such as gamma rays produced by radioactive decay processes in the soil and cosmic radiation originating from solar flares and other galactic objects, can ionize the air molecules and give rise to free electrons and positive and negative ions. Under normal conditions of temperature and pressure, the conduction in air at low field ranges from 10^{-16} to $10^{-17}\,A/cm^2$ in proximity to the Earth's surface, so normally air could be considered an excellent insulating material. When in an air-filled volume the electron concentration increases, an electrical breakdown process takes place and this gas becomes conductive.

1.2 IONIZATION PROCESSES IN GASES

As defined in IEEE Std. 539 (2005), the term "ionization" indicates "the process by which an atom or molecule receives enough energy (by collision with electrons, photons, etc.) to split it into one or more free electrons and a positive ion." This kind of collision is called *inelastic* because all the kinetic energy of the colliding particle or only a part of it is converted into potential energy of the atom or molecule. On the contrary, a collision is *elastic* when the total kinetic energy of the

Filamentary Ion Flow: Theory and Experiments, First Edition. Edited by Francesco Lattarulo and Vitantonio Amoruso.
© 2014 by The Institute of Electrical and Electronics Engineers, Inc. Published by 2014 John Wiley & Sons, Inc.

colliding particle is conserved during the collision. In this case the neutral atom or molecule does not acquire energy from the colliding particle and no excitation or ionization takes place. In a neutral gas at ambient temperature, almost all the collisions are elastic. On the contrary, the plasma formation includes several inelastic collisions (such as recombination, charge transfer, attachment, detachment, and dissociation) in addition to excitation and ionization of the atom. If an inelastic collision with an electron occurs, the neutral atom or molecule may become excited but not ionized by the acquired energy, so the same excited molecule may become ionized by a subsequent inelastic collision. The effectiveness of the ionization depends on the kinetic energy gained by the colliding electron along the free path between two successive collisions. Slow electrons or very fast electrons are poor ionizers because the amount of energy is insufficient for ionizing the atom or the period of interaction is too short in order to transfer the energy, respectively. Ionization processes may occur in either the absence or presence of an electric field. In the absence of an electric field, the behavior of electrons, atoms, and molecules inside a gas can be well represented by the kinetic theory. The macroscopic properties, such as pressure, temperature, distribution of particle velocities, and energies, are related to the average values of the velocities of the particles and of the mean free path. In the absence of an electric field, the heating of the gas at ordinary temperatures results in an equilibrium between the ionization and deionization processes. Gas heating at a very high temperature (thermal ionization) may change this equilibrium and promote an ionization process, as does application of a sufficiently high electric field. We will deal with the thermal ionization later on. Now we will consider the presence of an electric field.

Charge particles (electrons or ions) acquire kinetic energy from the field between each pair of collisions, so that the effectiveness of ionization depends on the amount of energy gained along the mean free paths λ in the direction of the applied electric field $\mathbf{E}$ and on its strength. All things considered, the average kinetic energy acquired by a particle carrying a charge q is $w_k = qE\lambda$. Incidentally, the velocity v given to the particle in the direction of an unitary field is called mobility: $k = v/E$.

The main processes considered here are:

- Ionization by electron and photon impact
- Attachment and detachment, which are important in electronegative gases
- Recombination

1.2.1 Ionization by Electron Impact

The ionization by electron impact is the principal collision process in electrical discharge phenomena. If a free electron acquires under the action of an applied field **E** a kinetic energy equal to or larger than the ionization energy w_i of the atom it collides with, an ionization process occurs and results in another electron and in a positive ion, which become accelerated and acquire kinetic energy. If the energy transferred to the atom during the collision is lower than w_i, then some electrons in the atom can acquire a quantum of energy and rise to a higher energy level without leaving the atom. The atom switches from a normal state to an excited state, and the quantum of energy required for this change is known as excitation energy w_e. These collisions are called *inelastic*.

Normally, the excited state of an atom lasts about 10^{-8} s. Moreover, some group II elements of the periodic table, as well as the inert gases, could present a metastable state—that is, a lifetime in some of their excited electronic states extended to seconds. This relatively high potential energy makes the metastables able to ionize neutral particles. When an electron drops to a lower energy level, the atom releases the quantum of energy in the form of a radiated photon having a wavelength $\lambda_p = h\,c/w_e$, where c is the free-space light velocity and h is the Planck constant. A radiated photon could transfer its energy to another atom or molecule, causing their excitation (photoexcitation) or ionization (photoionization) before vanishing. Note that a photon has neither mass nor electrical charge, which implies that it is unaffected by the background field. Positive ions result from an ionization process involving, for example, an energized electron and a neutral atom or molecule. The ion mass is comparable to the mass of the gas molecule, therefore each collision with other atoms causes a loss of about 50% of energy. Under the action of an applied field, the velocity acquired by the electron is largely greater than one acquired by the positive ion. The probability of a successive ionization determined by a positive ion is much lower than the ionization probability of an electron accelerated by the same applied field.

1.2.2   Townsend First Ionization Coefficient

The electric field distribution in the gas is the physical consequence of a voltage applied between two electrodes. Provided that the voltage is sufficiently high, a current will flow between the electrodes. Townsend (1910) first studied this current as a function of the applied voltage. In his experiment he used two conducting parallel plates located in a vacuum tube. An external ultraviolet radiation has been used to illuminate the cathode and to cause electron emission from the cathode surface. By reducing the pressure of the gas, it was possible to set the ions in motion at a sufficient velocity to ionize other molecules of the gas, even when the voltage employed has been of the order of magnitude of a few hundred volts. Townsend measured the current I

flowing across the plate gap as a function of the applied voltage V and found that the $I = I(V)$ curve initially increases linearly with the applied voltage. The rate of increase diminishes as the voltage increases further, and the current tends to attain a maximum value. This can be explained by considering that some electrons are emitted by the cathode and some positive ions diffuse in the tube and are lost by recombination on the tube walls. Diffusion, recombination, and losses on the walls decrease when the voltage increases until the voltage level exceeds, say, the value V_1 and all the electrons are collected on the anode. In a given region V_1–V_2 the current remains practically constant and approaches a saturation value. As the voltage increases further ($V > V_2$), there is a large increase in conductivity, presumably because new ions are produced by electron impact. The involved electrons gain sufficient energy from the electric field for ionizing the neutral gas molecules. The growth of the current in an even greater region V_2–V_3 obeys an exponential law, while in the extreme V_3–V_b region delimited by the breakdown voltage V_b the current increases faster than the exponential growth. This departure from the exponential law can be attributed to a secondary ionization process that generates new electrons (and then new avalanches) from the cathode by the impact of energized positive ions. In order to explain this current increase, Townsend defined an ionization coefficient α as the number of electron–ion pairs produced in the gas by a single electron which moves through a unit distance in the direction of the applied field. It is a matter of general knowledge that the coefficient α is referred to as Townsend's first ionization coefficient. This coefficient varies in function of the electric field for different gases. The related phenomenon is also called a primary ionizing process.

Townsend derived also the following expression:

$$\frac{\alpha}{p} = A_T \exp\left(\frac{-B_T}{(E/p)}\right) \tag{1.1}$$

for evaluating the primary ionization coefficient α in the function of the electric field. The constants A_T and B_T in Eq. (1.1) depend on the gas under examination and ambient conditions [see, for example, Table 1.2 in Abdel-Salam and Stanek (1988) where experimentally validated values of the coefficients A_T and B_T for air under standard conditions are reported]; p is the ambient pressure.

Additionally,

$$\frac{\alpha}{p} = A_T\left[\left(\frac{E}{E_0}\right)^2 - 1\right], \qquad E \geq E_0 \tag{1.2}$$

where E_0 is the critical electric field (namely, the voltage gradient on that confined surface area of an electrode on which a continuous corona is first detected when the applied voltage is gradually increased) and A_T is a constant [see Loeb (1939)]. Equations (1.1) and (1.2) are used in numerical procedures for simulating corona discharges (see Chapter 2).

1.2.3 Electron Avalanches

Consider a free electron positioned, say, on the cathode surface $(x = 0)$ under the influence of an uniform electric field. If the electron acquires enough energy, it can ionize a gas molecule by collision and produce another electron. These two electrons, in turn, acquire energy from the electric field and repeat the process. In this way, the number of electrons increases with x. Let n_0 be the initial number of electrons emitted from the cathode and let n_x be the increased number of electrons at distance x from the cathode. After denoting by α_e the number of ionizing collisions per unit length, made by an electron traveling along the x-direction of the electric field, the formula giving the elementary increase dn_x of the number of electrons increasing across the length dx can be written

$$dn_x = \alpha_e n_x dx \tag{1.3}$$

By integrating Eq. (1.3) across the cathode-to-anode distance d,

$$n_d = n_0 e^{\alpha_e d} \tag{1.4}$$

is obtained. The exponential growth $e^{\alpha_e d}$ of electrons represents the electron avalanche. In terms of current, if I_0 is the current leaving the cathode, then Eq. (1.4) becomes

$$I = I_0 e^{\alpha_e d} \tag{1.5}$$

Even though the distribution of charge carriers in an avalanche influences the applied uniform electric field E, Eq. (1.4) is obtained by assuming, in combination, that such an influence is negligible and the probability of an electron ionizing a gas molecule is constant and independent of the distance traveled in the electric-field direction. Because the velocity of electrons is larger than the velocity of positive ions, the electrons build up the head of the avalanche oriented toward the anode, whereas the positive ions form a long tail between the avalanche head and the cathode. Indeed, this charge distribution can significantly affect the local electric field.

1.2.4 Photoionization

A neutral gas atom A_0 colliding with an electron of energy lower than w_i may pass from the normal state to an excited state, according to the reaction $A_0 + e + w_k \rightarrow A^* + e$, with w_k and A^* respectively being the kinetic energy of the electron and the atom in an excited state. On recovering from the excited state within a short time ranging from about 10^{-10} to 10^{-7} s, the atom radiates a quantum of energy of a photon. Formally, $A^* \rightarrow A_0 + h\nu$ where the quantum of energy $h\nu$ in turn may ionize another atom whose ionization potential energy is equal to or lower than the photon energy. Accordingly, the process $A_0 + h\nu \rightarrow A^+ + e + w_k$ stands for photoionization. The excess of the quantum $h\nu$ over w_i may be converted into the kinetic energy of the released electron. Photoionization is a secondary ionization process and may act in Townsend's breakdown mechanism, in the streamer the breakdown mechanism, and in corona discharge. If $h\nu < w_i$, then the photon may be absorbed from the atom that passes to an excited state, so that the process is in this case called *photoexcitation*.

1.2.5 Other Ionization Processes

Ionization by Interaction of Metastables with Atoms   Some group II elements of the periodic table, as well as the inert gases, could present a metastable state—that is, a lifetime extended to a few seconds in some of their excited electronic states. This relatively high potential energy makes the metastables able to ionize neutral particles. If W^{**} is the energy of a metastable, the latter denoted by A^{**}, and w_{iB} the ionization energy of another atom B_0, then the following are the resulting reactions:

- If $W^{**} > w_{iB}$: $A^{**} + B_0 \rightarrow A_0 + B^+ + e$ (collision or Penning ionization)
- If $W^{**} < w_{iB}$: $A^{**} + B_0 \rightarrow A_0 + B^*$ (excitation of atom B_0)

Denoting with w_{iA} the ionization energy of the atom A, in the case of high-density metastables, and provided that $2W^{**} > w_{iA}$,

$$A^{**} + A^{**} \rightarrow A_0 + A^+ + e + w_k$$

is the presumed reaction. Moreover, a metastable atom may be ionized by photon absorption

$$A^{**} + h\nu \rightarrow A^+ + e$$

before returning to its ground state.

Thermal Ionization   The velocities of the atoms of a gas in thermal equilibrium are distributed according to Boltzmann's distribution. If a gas is heated to sufficiently high temperature, many of the gas atoms acquire sufficiently high velocity to cause ionization by collision with other atoms or molecules. Thermal ionization is the principal source of ionization in flames and high-pressure arcs. Saha (1920) claimed that the thermal ionization in air is significant only at temperatures above about 4000 K. Thermal-ionization reactions are of particular importance for describing the behavior of plasmas created via electric discharge, such as arc discharges.

Autoionization   This is a process by which atoms or molecules spontaneously emit one of the shell electrons so that the atom passes from an electrically neutral state (metastable) to a singly ionized state.

1.3   DEIONIZATION PROCESSES IN GASES

1.3.1   Deionization by Recombination

When groups of positively and negatively charged particles coexist in a gas, processes of recombination

$$A^+ + B^- \rightarrow A_0 B_0 + h\nu$$

may take place. Here, B^- may be an electron or a negative ion. Radiative recombination, which may be considered as the reverse of photoionization, occurs only when electrons are involved. Recombination of positive and negative ions consists of two phases: In the first phase, two ions in random motion perform elliptic or hyperbolic orbits, under the action of Coulomb forces, around their common center of masses. In the second phase, a charge transfer takes place during the orbital encounter with consequent charge neutralization. The difference between the ionization energy of the positive ion and the electron affinity of the negative ion gives rise to a kinetic energy increase for the neutralized particle.

1.3.2   Deionization by Attachment

After a number of collisions with neutral atoms, or when the field strength is reduced, the kinetic energy of the electron becomes insufficient to excite the atom and then the electron could get attached to the atom to form a negative ion. Some atoms or molecules lacking one or two electrons in their outer orbits tend to capture free electrons to become negative ions. This process forming a negative ion is known as *electron attachment* ($A_0 + e \rightarrow A^-$). Gases prone to exhibit such a behavior are called *electronegative gases*, and the energy required to remove an electron from a negative ion for restoring neutrality is called the *electron affinity* of the atom. The attachment of electrons to neutral molecules may be taken into account by an attachment coefficient η (analogous to the ionization coefficient α), thus representing the number of negative ions created by a single electron moving through a unit distance in the field direction.

Some elements of the periodic table (for example, F, Cl, Br, O, S) are lacking one or two electrons in their outer shell and tend readily to acquire a free electron to form a stable negative ion. This physical property is called *electronegativity*. The so-formed negative ion remains stable until its total energy is lower than that of the neutral atom. The excess energy upon attachment can be released as a radiative photon

$$e + A_0 \rightleftharpoons A^- + h\nu$$

or as the kinetic energy of a colliding third body

$$e + A_0 + B_0 \rightleftharpoons A^- + (B_0 + w_k)$$

The reverse process, namely, the electron detachment, requires energy, termed the electron affinity of the atom, for removing the electron and restoring neutrality.

Other processes of negative ion formation are:

- *Dissociative attachment*, in which the excess energy is used to separate the molecule into a neutral particle and an atomic negative ion:

$$e + A_0 B_0 \rightleftharpoons A^- + B_0$$

   Alternatively, the excess energy is released to a colliding particle as kinetic and/or potential energy after forming the molecular negative ion:

$$e + A_0 B_0 + A_0 \rightleftharpoons (AB)^- + A_0 + w_k + w_p$$

- *Splitting* of a gas molecule into positive and negative ions upon impact of an electron without attachment:

$$e + A_0 B_0 \rightleftharpoons A^- + B^+ + e$$

- *Charge transfer* following a heavy particle collision:

$$A_0 + B_0 \rightarrow A^+ + B^-$$

Two attachment processes can occur in air. For electron energies ranging between 0.2 and 0.5 eV, molecular ions could be formed according to the reaction

$$O_2 + O_2 + e = O_2^- + O_2$$

For electron energies equal to about 2.9 eV, atomic ions are formed according to

$$O_2 + e = O^- + O$$

The above attachment processes involving electrons may be expressed by a relation similar to Eq. (1.5) describing electron avalanche in a gas. The formula

$$dI = -\eta I dx \qquad (1.6)$$

governs the elementary electron current loss on a distance dx. For a gap of length d with a current I_0 starting at the cathode, the integral version of Eq. (1.6) becomes

$$I = I_0 e^{-\eta d} \qquad (1.7)$$

1.4 IONIZATION AND ATTACHMENT COEFFICIENTS

Some results in Harrison and Geballe (1953) showed that the ionization α and attachment η coefficients in oxygen and in air are comparable in the range $25 \leq E/p \leq 60 \text{ V cm}^{-1} \text{ Torr}^{-1}$. By considering only the electron collision and electron attachment processes, the resulting number of free electrons on a distance dx is given by

$$dn_x = n_x(\alpha - \eta)\, dx$$

Integrating from $x = 0$ to x, with n_0 electrons starting from the cathode, gives

$$n_x = n_0 e^{(\alpha - \eta)x}$$

namely, the number of electrons at any point x in the gap. The quantity $(\alpha - \eta)$ in the exponent is denoted by $\bar{\alpha}$ and is acknowledged as the effective ionization coefficient. As a consequence, the total steady-state current is the combinative result of a flow of electrons and negative ions. A given increase of negative ions on the distance dx can be written

$$dn_- = n_x \eta \, dx = n_0 \eta e^{\bar{\alpha}x} \, dx$$

Performing the integration in the limits 0 and x gives

$$n_- = \frac{n_0 \eta}{\bar{\alpha}} [e^{\bar{\alpha}x} - 1]$$

so that

$$I = I_0 \left[\frac{\alpha}{\bar{\alpha}} e^{\bar{\alpha}d} - \frac{\eta}{\bar{\alpha}} \right] \tag{1.8}$$

is ultimately obtained for the total current after summing the two components n_x and n_-. It is simple to verify that Eq. (1.8) reduces to $I = I_0 e^{\alpha d}$ in the absence of attachment ($\eta = 0$).

Indeed, the current measurement between parallel plane electrodes shows that the rate of increase of the current I is higher than the one given by Eqs. (1.7) or (1.8) as the voltage is significantly increased. To explain this discrepancy, Townsend postulated that an overlooked subsidiary mechanism plays a significant role in affecting the current. He first considered a release of electrons in gas by a collision of positive ions and, later, a release of electrons from the cathode by positive ion bombardment. Other processes responsible for the raised departure include secondary electron emission at the cathode by photon impact and photoionization of the gas itself. Townsend considered a second coefficient γ, referred to as "the secondary ionization coefficient," denoting the number of electrons released from the cathode because of a subsidiary emission mechanism.

1.5   ELECTRICAL BREAKDOWN OF GASES

The previously described avalanche process is important in order for a breakdown mechanism to develop. Two typical breakdown mechanisms, each of them operating under specifically favorable conditions, are discerned and labeled Townsend's mechanism and the streamer mechanism. The former takes place when the product of pressure and electrode spacing (in a uniform gap) does not exceed about 5 bar mm. If this limit is exceeded, the space charge of the avalanche could be large enough to significantly change the background field, a circumstance responsible for an avalanche-to-streamer transition. The latter mechanism is extensively treated in Section 1.6.

1.5.1 Breakdown in Steady Uniform Field: Townsend's Breakdown Mechanism

Let n_c be the total number of electrons emitted from the cathode surface and let n_0' be the number of electrons emitted by secondary ionization processes, so that $n_c = n_0 + n_0'$. The average number of collisions produced in the gap by each electron leaving the cathode is given by $(e^{\alpha d} - 1)$. Hence, the number of ionizing collisions in the gap is $n_c(e^{\alpha d} - 1)$. If γ denotes the efficiency of the secondary electrons' emission process, then the number of secondary electrons is

$$n_0' = \gamma n_c \left(e^{\alpha d} - 1\right) = n_c - n_0$$

from which

$$n_c = \frac{n_0}{1 - \gamma(e^{\alpha d} - 1)}$$

The number of electrons attaining the anode is

$$n_a = \frac{n_0 e^{\alpha d}}{1 - \gamma(e^{\alpha d} - 1)}$$

and the steady-state current becomes

$$I = \frac{I_0 e^{\alpha d}}{1 - \gamma(e^{\alpha d} - 1)} \tag{1.9}$$

At lower field strengths $e^{\alpha d} \rightarrow 1$, thus implying $I = I_0 e^{\alpha d}$ (exponential law in the V_2–V_3 region). As V increases, both $e^{\alpha d}$ and $\gamma e^{\alpha d}$ increase in such a way that $e^{\alpha d} \gg 1$ and $\gamma e^{\alpha d}$ close to 1 are expected occurrences. Therefore, the current I approaches infinity because

$$\gamma(e^{\alpha d} - 1) = 1 \tag{1.10}$$

under the described circumstances. Such a condition is known as Townsend's criterion for breakdown. An alternative expression for Eq. (1.10) is

$$\alpha d = \ln\left(1 + \frac{1}{\gamma}\right) = K'$$

Because the value of γ is very small ($<10^{-2}$–10^{-3}), the above logarithm ranges from 8 to 10 in a Townsend's discharge. The exponential quantity $e^{K'}$

expresses the number of electrons forming the avalanche head that gives rise to the avalanche-to-streamer transition. In deriving Eq. (1.9), only the impact of positive ions on the cathode has been considered as a secondary ionization process. Further secondary processes may be considered, namely, gas ionization by positive ions (according to Townsend's original assumption), photo-emission from the electrode, collision of metastable ions on the cathode, and gas ionization by photons. Accordingly, Eq. (1.9) becomes

$$I = \frac{I_0 e^{\alpha d}}{1 - \gamma_p \left(e^{\alpha_p d} - 1 \right)} \tag{1.11}$$

where γ_p and α_p assume specialized expressions.

1.5.2 Paschen's Law

Consider the simple electrode assembly represented by a pair of parallel plates a distance d apart, where a gas at pressure p is the filling medium. By increasing the voltage V applied between the electrodes, an electrical breakdown occurs as soon as the value $V = V_b$ is reached. Paschen's law expresses the experimental evidence that the critical value V_b, termed the breakdown voltage, is a function of the product pd. The given curves show a minimum at $(pd)_{\min} = 10^{-2}$ bar mm, or thereabout (the minimum differs from one gas to another), and the corresponding electric field is E_m. Under the described conditions, an electron crossing the gap will produce a certain number of ionizing collisions. If $pd > (pd)_{\min}$, the number of collisions made by an electron increases and hence the energy lost in collisions is higher than one at $(pd)_{\min}$. Therefore, the probability of ionization decreases as long as E_m remains unchanged, or, differently speaking, the electric field must be increased for this loss to be compensated. Conversely, if $pd < (pd)_{\min}$, then the number of collisions and, hence, the number of ionizing collisions decreases with respect to the value at $(pd)_{\min}$. In this case, the increased ionization probability after each collision can be achieved only by increasing the energy gained by electrons within a mean free path. The consequence is that an electric field higher than the previous value of E_m is required.

The validity of Paschen's law has been confirmed experimentally, provided that the limit temperature 1100°C is not surpassed [see Alston (1968)]. It is ensured that above 2000 K, Paschen's law fails because the thermal ionization no longer can be neglected. Moreover, in order to take into account the effect of the temperature, Paschen's law must be expressed as a function of the product $(\rho_m d)$ where ρ_m is the gas density [see Abdel-Salam (1976)]. High pressure gives rise to a departure from Paschen's law because of the predominant role assumed by the field emission. Some empirical relations have been suggested by Ritz (1932), Holzer (1932), Boyd, Bruce, and Tedford (1966), and Alston (1968) to express the breakdown voltage V_b [kV] of the uniform field in air gaps of width d [m] at atmospheric pressure. These substantially can be formulated as follows:

$$V_b = A_1 d + A_2 d^{1/2}$$

and an excellent agreement with experiment is given in Boyd et al. (1966) for $A_1 = 2449$ and $A_2 = 66.1$ (average and maximum errors equal to 0.675% and 1.83%, respectively), even though A_1 and A_2 are deprived of physical meaning. Alternative formulas are

$$V_b = 6.04\sqrt{pd} + 23.91(pd)$$

given by Bruce (1953), with p expressed in kPa, which is applicable for pd values within the limits of the validity of Paschen's law, and

$$V_b = \frac{Bpd}{\ln(100pd) + B_1}$$

provided by Townsend (1910), where $B_1 = \ln\{D/(\ln(1 + \gamma^{-1}))\}$. In this case, physical meaning can be ascribed to B and D because they are derived from Townsend's formula [see Eq. (1.4)] involving the first ionization coefficient. Therefore, D is acknowledged to be the saturation ionization in the gas at high values of E/p, while $B/2$ stands for the value of E/p at the inflection point of the α/p versus E/p curve (see Section 1.2.2) with which the paper by Heylen (1973) is concerned. It is ensured there that imposing the values of 44,804 and 14.49 to the constants B and B_1, respectively, gives a better fitting with the experimental values under normal pressure conditions (the average error is 0.485% with a maximum error of 0.85%).

1.6 STREAMER MECHANISM

Townsend's mechanism attempts to explain the formation of spark breakdown in uniform field gaps as a series of successive avalanches, but the spark breakdown's time lags obtained theoretically are not consistent with the very shorter values observed by experiment. Moreover, Townsend's mechanism applied to long gaps is not able to account for the branched configuration and irregular growth of the channel. A streamer theory applied to spark breakdown was proposed by Loeb and Meek (1941) for positive streamers and by

Raether (1956) for negative streamers. Both versions assume that the spark discharge comes from a single avalanche (whose space charge develops a plasma streamer), the electric field is locally enhanced by the space charge (notably, on the avalanche head), and the photoionization of gas molecules surrounding the avalanche is the crucial phenomenon for developing a self-propagating streamer. Loeb's version of the streamer theory accounts for the cathode-directed streamer formation by assuming that a first avalanche crosses completely the interelectrode gap (in an uniform field) and leaves behind it a cone-shaped volume of positive ions. Photoelectrons are produced in the gas surrounding the avalanche and initiate auxiliary avalanches that are oriented along the direction of the total field composed of the space-charge field and the exogenous uniform field. The process initiates near the anode where the local space-charge field is higher. The negative charge of the head of such auxiliary avalanches is attracted by the positive cone-shaped volume, and several positive branches (tails left by the auxiliary avalanches), termed streamers, intensify the surrounding field toward the cathode. The process, which repeats and develops from the anode toward the cathode, results in the formation of a conducting filament of highly ionized gas bridging the gap, throughout. Raether (1964) accounts for the negative anode-directed streamer formation by assuming that streamers develop when the initial avalanche produces a sufficient number of electrons, namely, such that the given space-charge field is comparable to the applied field. The enhanced total field is the primary cause giving rise to a number of photoionization promoted, secondary anode-directed avalanches ahead of the initial one, thus forming a negative streamer.

1.7 BREAKDOWN IN NONUNIFORM DC FIELD

A Townsend-type equation governing current growth in a nonuniform field due to primary and secondary ionizing processes is available in Pedersen (1989). Let N_c be the total number of primary ionizing collisions in the gas per primary electron emitted from a small surface of the cathode. If I_0 is the electron current due to the external source emitted from that cathode area, then

$$I = I_0(1 + N_c)$$

is the discharge current only due to the primary ionizing process. Here, N_c is the total number of primary ionizing collisions in the gas per primary electron emitted from the cathode. This discharge current will develop along a line of force from the cathode to the anode. The secondary ionizing processes, subject to a higher onset level, can be described by the number M_e of secondary electrons released at the cathode per emitted primary electron. The consequence is that the formula for the discharge current produced by succeeding generations of avalanches reads

$$I = I_0(1 + N_c)(1 + M_e + M_e^2 + \cdots) \tag{1.12}$$

which approximately becomes

$$I = I_0 \frac{(1 + N_c)}{(1 - M_e)} \tag{1.13}$$

for $M_e < 1$. Under such conditions, the current assumes a finite value, whereas for $M_e \geq 1$ Eq. (1.12) diverges, thus predicting breakdown.

To determine the parameter N_c, pay attention to the generic distance x from the cathode and to that, expressed with the notation d, between the cathode and anode, both measured along a line of force of the electric field. The number of ionizing collisions between x and $x + dx$ due to a primary electron is then given by

$$dN_c = N_e(x)\alpha \, dx$$

where $N_e(x)$ is the number of electrons at distance x per primary electron (or the total number of electrons, residing in the head of the electron avalanche, which turns out to be a surrogate of the avalanche size) obeying the law

$$N_c(x) = \exp\left[\int_0^x (\alpha - \eta) \, dx \right]$$

Therefore,

$$N_c = \int_0^d \exp\left[\int_0^x (\alpha - \eta) \, dx \right] \alpha \, dx$$

The parameter M_e is related to N_c by the equality $M_e = \gamma N_c$, with γ also termed the Townsend secondary ionization coefficient. As a result, Eq. (1.13) becomes

$$I = I_0 \frac{\left(1 + \int_0^d \exp\left[\int_0^x (\alpha - \eta) \, dx \right] \alpha \, dx \right)}{\left(1 - \gamma \int_0^d \exp\left[\int_0^x (\alpha - \eta) \, dx \right] \alpha \, dx \right)}$$

and, then,

$$\gamma \int_0^d \exp\left[\int_0^x (\alpha - \eta)\, dx\right] \alpha\, dx = 1 \qquad (1.14)$$

which results in the criterion for Townsend's mechanism of spark break-down in an electronegative gas under nonuniform fields. A succession of electron avalanches, initiating to the cathode, is the prerequisite for Townsend's breakdown mechanism to develop. However, Townsend's criterion is hard to manage for engineering problems because the secondary ionization coefficient γ is very sensitive to the actual conditions of electrode surfaces and gas purity. Furthermore, measurements of γ have been obtained up to now only at technically unimportant pressure values of lower than 3.4 kPa. Townsend's criterion is also unable to explain breakdown under steep voltage surges.

1.8 OTHER STREAMER CRITERIA

The streamer mechanism assumes that the growth of a single electron avalanche becomes unstable before reaching the anode, even though fast-moving streamers from the avalanche head are permitted by photoionization. These streamers form a highly conducting channel across the gap, thus causing a definitive voltage collapse. Both Meek and Cragg (1953) and Reather (1964) independently developed a new equation for the breakdown of a nonuniform field gap, and Reather additionally suggested that the critical number of charge carriers for the avalanche-to-streamer transition is in the 10^8 range, irrespective of the gas pressure. By modifying Meek, Cragg, and Reather's equation, Pedersen proposed a semiempirical streamer criterion according to which

$$\int_0^x (\alpha - \eta)\, dx = \ln(N_c) = \text{const} \qquad (1.15)$$

Here, x is that fraction of the gap length d representing the distance between the cathode and the point where $\alpha = \eta$, and N_c is the critical size of the avalanche. When $\alpha > \eta$, the electron avalanche grows, whereas the contrary occurs for $\alpha < \eta$. Both α and η are functions of E/p, so that in a uniform E-field there is a limiting field strength, one related to the condition $\alpha = \eta$ below which breakdown cannot occur because all the electrons are attached. This effect has been verified experimentally by Geballe and Reeves (1953), Crowe and Devins (1956), and Bhalla and Craggs (1962). The values of the constant k, which depend on the product *pd* and on the nature of the electronegative gas, have been computed by Malik (1981) as a function of the applied uniform electric field. For air at $pd = 1$ bar cm, the constant in Eq. (1.15) is equal to 18.

1.9 CORONA DISCHARGE IN AIR

The term *corona* evokes partial discharges in air which, owing to physical and technological reasons limiting the applied voltage, can burn on the over-stressed zones of hot electrodes. This kind of discharge occupies a short layer (the ionization region), attached to the energized conductor, in comparison to the outer region of the gap crossed by drifting charges. In general, there are a number of active zones that are individually very confined and collectively disseminated over the electrode surfaces where the local curvature increases so much that the enhanced electric field can trespass there an established onset value. Owing to several wanted or unwanted effects concerned with corona discharges, the overall performances of this complex phenomenon have extensively been detected in the laboratory and described by theoretical models. The investigation is aimed at

- understanding the dependence of corona activity from several physical parameters,
- paying additional attention to undesired companion effects such as radio noise, power loss, and audible noise, and
- designing processes and devices for practical applications,

to name a few objects. The geometrical configurations adopted are very different. As a result, many of the modes of the corona appearance have been reproduced during experimental tests as a function of the applied voltage's polarity and magnitude, field divergence, protrusion's form and surface distribution, ambient conditions, and so on. The next subsection summarizes the characteristics of the various coronas given.

1.9.1 DC Corona Modes

Because of the low mobility of ions, space charges of both polarities accumulate in the gap near the stressed electrode. As a result, positive ions are formed by an ionization process, and negative ions are formed by an attachment process favored by the electronegative properties of air. The resulting space charge causes distortion of the local field and, ultimately, discharge. Corona modes differ according to the equilibrium between two opposite situations that are identified as creations of ionic space charge and electrostatic removal.

1.9.2   Negative Corona Modes

When the highly stressed electrode is the cathode, electron avalanches develop from the cathode toward the anode in a decreasing field. The ionization proceeds up to a distance from the cathode where $\bar{\alpha} = 0$. All the points characterized by such an equality form a boundary surface, say, S_0, where the avalanches stop. The tail of positive ions takes place between the cathode and S_0. The electrons at reduced mobility continue to migrate toward the anode and attach the oxygen molecules to form negative ions that, because of their slow drift velocity, accumulate in the gap beyond S_0. The so-formed two regions of space charge disturb the background electrostatic field, and a new field distribution enlivens the gap throughout. Substantially, the ionic space charge increases the field near the cathode and reduces the field near the anode. The surface S_0 is displaced toward the cathode, and the successive electron avalanches develop on a shorter distance where higher field strengths are present. Increasing the applied voltage leads to three modes of corona discharge, that is,

- Trichel streamers
- Pulseless glow
- Negative streamers

Trichel Streamers   While studying the negative corona discharge in air for a point-to-plane configuration, Trichel (1938) and Loeb (1965) observed that the corona current is composed by discrete pulses whose magnitude and frequency are functions of the applied voltage (provided that this is only slightly larger than the onset value), point size, and ambient conditions. A Trichel discharge propagates for a few tens of nanoseconds and is randomly distributed on the surface of the stressed protrusion. The distribution simultaneously involves several spots changing their position during the observation time. The related current oscillogram shows sharp peaks of very short duration (tens of nanoseconds with a rise time of about 1.3 ns) separated by longer interpulse intervals (tens of microseconds). The amplitude of these pulses, being of the order of 10^{-8} A at a thin point electrode up to a few tens of milliamperes for a larger electrode, decreases as the applied voltage [see Trih and Jordan (1968)], pressure, and humidity [Bian et al. (2009)] increase. On the contrary, the pulse frequency increases as the voltage increases. The pulse frequency is a function of the time spent to remove the ionic space charge through the drift under the action of the applied field. Trichel's pulse frequency can be increased up to a critical value depending on the electrode geometry, pulse amplitude, and surface conditions. For a sphere of 8-mm diameter, the critical frequency is 2 kHz. Trichel's pulse converts into a new corona mode when any critical frequency is reached.

The pulsating nature of the Trichel streamer can be explained by taking into account the interaction between the ion space charge and applied field. Once a Trichel streamer is initiated and the discharge is self-sustained by the photon-induced secondary electron emission, two oppositely charged ion clouds are formed and localize in the gap in such a way as to influence the electric field and force the boundary surface S_0 to settle down in closer proximity to the cathode. The positive ion cloud is partially neutralized at the cathode by the negative ions produced by successive avalanches. The residual negative ionic charge reduces the field intensity at the cathode below the onset field and then the discharge is suppressed. This phase is followed by a period, the so-called dead time, during which the remaining negative charge is dispersed by the field and a new streamer will develop when the ionic charge has been sufficiently removed by the drift. The streamer frequency depends on the ion-removal velocity and increases with the growing applied voltage up to a critical value of the frequency: At some high fields, the pulse repetition rate can decrease and a transition to a new corona mode can occur.

Pulseless Glow By increasing the voltage even further until a critical frequency is obtained, the transition from Trichel's pulses to pulseless glow occurs. The discharge becomes stably positioned on the protrusion surface and, in a visible sense, it shows the basic features of the glow discharge; that is, a cathode dark space is followed, respectively, by a bright negative glow, a Faraday dark space, and a luminous conical positive column. During this corona mode, the electrons are emitted at low kinetic energy from the cathode (as a result of ionic bombardment) and acquire energy from the field while crossing the cathode dark zone. The consequent intensive ionization, occurring in the negative glow region, is detrimental for the electron kinetic energy. Going beyond the periphery of the negative glow region, the electrons are once again accelerated across the Faraday dark zone. The successive positive column is the result of ionization of gas atoms. The conical shape of this positive column is attributed to the diffusion of free electrons in the low-field region. The detected discharge stability in this corona mode is attributable to higher applied voltages because the field becomes more efficient in removing the negative ionic space charge and, consequently, in warding off suppression of the ionization activity. The corona current is stable in time and increases with the voltage until this mode of discharge turns into negative streamers. The reversible character of the mode transitions depends, of course, on the generally linear mechanisms involved in the discharge formation and extinction.

Negative Streamers The last mode of corona discharge at the cathode is represented by negative streamers. These occur when the voltage is increased still further. This corona mode shares several aspects with the previous pulseless glow, along with the fact that the conical positive column is forced to form a streamer stem extending farther into the gap with no branching. The length of the streamers increases with the voltage until one of them crosses the gap, causing breakdown, or approaches a streamer eventually initiated from the anode. This corona mode also depends on electron emissions from the cathode as a result of ionic bombardment, while the formation of a streamer channel characterized by intense ionization is indicative of even more effective space-charge removal under the action of the applied field. The streamer current consists of a DC component, with superimposed pulses, which allows the discharge to never disappear altogether. The corona current increases continuously with the voltage until the discharge mode switches in proximity to the breakdown region.

1.9.3 Positive Corona Modes

When the polarity of the highly stressed electrode is positive, each electron avalanche develops from the boundary surface S_0 (where $\bar{\alpha} = 0$) toward the anode in a continuously increasing field. These favorable conditions for avalanche formation result in a critical field intensity that is slightly lower for positive than for negative coronas. A positive ion space charge (streamer) is left along the avalanche path in consequence to the lower mobility of the ions. Meanwhile this positive streamer is moving away from the point of the electrode, and photoelectrons start from the streamer and cause ionization with the consequent formation of variously directed avalanches in proximity to the electrode surface. As a result, a discharge spreads out over the surface points. The free electrons formed lose their energy by ionization of the neutral molecules near the anode, and then they can be neutralized at the anode or recombine with positive ions or create negative ions by attachment. The consequent presence of space charge of both polarities causes field reduction in the region close to the surface and enhancement in the farther region. This happens where secondary electron avalanches may be attracted to promote outwardly directed propagation of the discharge, along a streamer channel, into the gap. Four positive corona discharge modes can be observed, as the electric field at the anode is increased, prior to a definitive breakdown. These modes are summarized as follows in the order of appearance:

- Burst corona
- Onset streamers
- Positive-glow discharge
- Breakdown streamers

Burst Corona This corona mode displays a thin, luminous sheath close to the anode as a visible result of ionization spreading and avalanche formation in proximity to the anode surface. The positive ions created and left near the anode tend to reduce the field with consequent discharge suppression. The resulting discharging current consists of very small positive pulses.

Onset Streamers The positive ion charge left by the first avalanche enhances the field and produces a number of photons emitted in all directions. Some air molecules will be photoionized, and each photoelectron produced within the ionization zone will be accelerated under the action of the resultant electric field. New avalanches (called "second generation") will develop, and the discharge will assume an elongated aspect with several filamentary channels stemming from it. The absorption of free electrons at the anode gives rise to a residual positive charge that reduces the local field and ultimately causes streamer-discharge suppression. A dead time is required for the applied field to remove the ionic charge and restore the conditions for new streamer development. Thus the discharge current is a pulse of short duration, high amplitude, and relatively low repetition rate. The onset-streamer current ranges from a few tenths of milliamperes in highly divergent fields to a few hundreds of milliamperes when larger electrodes are involved. In the case of an 8-mm-diameter sphere protruding from the curved surface of a cylindrical conductor, the measured currents revolved around 250 mA. With reference to a conical boss with cone angle 30°, the current was about 3 mA. The mean rise time of the pulses was of the order of 30 ns and their half-peak time of about 100 ns [see Trinh and Jordan (1968)].

Burst corona and onset streamers develop in an alternative way over a small range of voltages immediately beyond the corona onset. In fact, for these voltage levels, the increased field is more effective in removing the positive space charge in close proximity to the anode surface, thus promoting a side-burst corona at the anode. A few microseconds after the suppression of the streamer, the anodic region is so rapidly cleared of the positive ionic space charge that incoming negative ions encounter a sufficiently high field to shed an electron at instant of impact. This free electron sustains ionization activity over the anode surface in the form of a burst corona, which develops until it is suppressed because of its own positive space charge. As the voltage is increased further, the space-charge removal at the anode becomes more effective and the burst corona enhances accordingly. As the burst corona develops, the positive ions created in the meantime are rapidly pushed away from the anode and accumulate in front of the anode. Once stably formed, the positive ionic space charge prevents radial development of discharge into the gap and a burst corona can more readily develop. This happens at the expense of the onset streamer until its suppression is accomplished. A new mode, termed positive-glow discharge (or Hermstein's glow), is then established at the anode.

Positive-Glow Discharge This corona mode consists of an intense ionization activity in the zones immediately adjacent to the anode and appears in the form of a thin, luminous layer adhering to the electrode surface. This activity is promoted by the rapid removal of positive space charge operated by the intense field. Moreover, the field intensity fails to such a great degree in allowing radial development of the discharge and streamer formation that the density of negative space charge becomes high enough to fully suppress onset streamers. As a result, this corona mode appears as a stable and uniform glow near the anode surface. The role assumed by the negative ions in forming such a stable glow was presumptively claimed by Trichel (1938) but experimentally proved by Hermstein (1960). That is why Loeb successfully proposed to call "Hermstein's glow" a positive glow. This emits discrete pulses of light and the resulting discharge current takes a saw-toothed waveform composed by a direct current (growing with the voltage) with superimposed small but stable pulses. The frequency of these burst pulses increases with the voltage and may attain a few megahertz. The pulsing nature of the glow corona becomes the subject matter carefully treated by Beattie (1975), Sigmond (1978), and Cross and Beattie (1980). In accord with these investigations, photo-detachment of negative ions by ion–electron recombination radiation provides seed electrons for pulse promotion. The transition from burst pulses to a stable glow seems to be more gradual when corpulent electrodes are adopted, in which case, however, the property of Hermstein's glow of suppressing streamers is less effective with respect to slender and pointed conductors.

Breakdown Streamers As the voltage increases, the stable positive glow loses its uniformity and one or two zones of higher luminosity are then formed. These spots of enhanced ionizing activity can be seen moving slowly over the anode and giving rise to breakdown streamers. The discharge parameters are of the same order of the onset streamers, but their elongation into the gap is larger and asymmetrical. The length, amplitude, and repetition rate of the pulses grow with the voltage. When the applied field becomes sufficiently high to remove the positive space charge from the anode region, the radial development of the discharge becomes feasible, which may result in a breakdown streamer.

1.10   AC CORONA

When an electrode is raised to a high sinusoidal potential, different corona modes of both polarities may be observed during a complete cycle depending on the applied voltage, gap length, and onset gradient. Contrary to the case of a short gap, in which the ionic space charge created during a half-cycle can impact a collector during the same half-cycle, the more usual case of a long gap is that in which it is larger than the maximum distance that the ionic space charge travels during one half-cycle. Under the circumstances described, the ionic space charge, created during one half-cycle, is drawn back toward the highly stressed electrode during the following half-cycle, thus before impacting the collector. As a consequence, the discharge development happens to be influenced in the sense that onset streamers are suppressed in favor of the positive glow discharge. Even negative streamers cannot be observed under AC voltage, because their onset field is greater than the breakdown voltage during the positive half-cycle. All things considered, only the following corona modes can now be categorized, namely, negative Trichel streamers, negative glow discharge, positive glow discharge, and positive breakdown streamers. When the applied voltage slightly exceeds the corona onset level, then negative Trichel streamers, positive onset streamers, and burst corona can be observed in the two half-cycles.

1.11 KAPTZOV'S HYPOTHESIS

When a conductor is in corona, then the electric field on the surface is somehow influenced by the surrounding ion space charge. Assessing the actual field strength on the active surface is of prominent importance in theoretical analyses subject to boundary conditions. Kaptzov's hypothesis (KH) consists in assuming that the space charge emitted into the interelectrode gap is in amounts that hold the surface field at the onset level. A theoretical calculation of the surface electric field for positive and negative polarities is extensively available elsewhere [see Khalifa and Abdel-Salam (1973)], whereas some experimental tests prove that the surface field is lower than the corona onset level [see Waters, Rickard, and Stark (1972)]. Provided that this is the case, KH would also invalidate to some extent the theoretical prediction of current losses, in the sense that the true corona-originated losses would be greater than those given by calculations.

For cylindrical conductors, the corona onset field is generally calculated by Peek's formula

$$E_0 = E_{0p} m \delta \left(1 + \frac{K}{\sqrt{\delta r}} \right) \qquad (1.16)$$

where E_0 [kV/cm] is the corona onset gradient; r [cm] is the conductor radius; $\delta = \frac{3.92}{273+t} p$ is the air relative density (pressure p in cmHg and temperature t in °C); m is the conductor surface's irregularity factor (ranging from 0.8 to 1); $E_{0p} = 29.8$ kV/cm and $K = 0.301$ in AC and in DC (negative polarity) cases for two parallel conductors above ground; and $E_{0p} = 33.7$ kV/cm and $K = 0.24$ in the DC case (positive polarity) for two parallel conductors above ground.

CHAPTER 2

ION-FLOW MODELS: A REVIEW

2.1 INTRODUCTION

Modeling a unipolar ion flow is a difficult task that has been shared by a number of researchers in the past century. This issue is still being pursued nowadays because, apart from some theoretical difficulties left unresolved (see, in particular, Chapter 4), broad avenues are opening up in applied electrostatics. According to a simple representation, ions already generated at, and with same polarity of, the electrode under corona (often also referred to as the emitter or source) are definitively subject to slow drift, toward a usually grounded counterelectrode (the collector), under the action of a total electric field $\mathbf{E}$. The space distribution of the current density turns out to be a meaningful descriptor, among others, of ion flows. Such a charge is essentially composed of two classes of carriers that are identified, owing to their own mobility k and nature, as small air ions and charged aerosols. Small air ions originate from air molecules that have been ionized during the corona process. These ionized molecules are subject to such a large number of collisions with other air molecules that in a few milliseconds they evolve into charged molecular clusters. These clusters, termed small air ions, consist of 3 to 10 water molecules electrostatically clustered around a kernel ion. When the charge of the molecular cluster is transferred or neutralized, the small air ion dissociates into its constituent molecules. Small air ions have a diameter of

Filamentary Ion Flow: Theory and Experiments, First Edition. Edited by Francesco Lattarulo and Vitantonio Amoruso.

about 1 nm. When a small ion collides with an aerosol (suspended liquid or solid particles categorized as aerosols exist naturally in ambient air), then the ion charge is transferred to the latter. Charged aerosols are much larger (diameter ranging from 0.015 to 0.1 μm) than small air ions and, consequently, are characterized by a comparatively lower mobility in the 3×10^{-4} to 4×10^{-3} cm^2/V·s range ($k \sim 1$ cm^2/V·s for small ions). Ions with a diameter of 1–10 nm are sometimes referred to as medium-sized ions or intermediate ions. In spite of the different ionic species involved, a constant average mobility for each polarity is usually assumed in the study of ion drifts (Chapter 4).

Most referential electrode arrangements for corona-originated unipolar ion-flow investigations are those assuming the wire-plane, point-to-plane, and coaxial (or, more in general, paraxial) configurations (see Chapter 5, Figure 5.1). A prominent practical example is represented by an HVDC transmission line, typically regarded as a composed version of the referential wire-plane assembly. In any case, the corona-originated ions with the same polarity of the source (active conductor) are those that drift to the collector, thus filling in any given instant (say, of a steady state) the source-to-collector spacing. Theoretical and experimental investigations on the subject point toward different objects, typically including the evaluation of corona losses, audible noise, and distribution laws of the ionized field parameters, notably, surface distributions of electric field and ion current at a grounded planar collector. Specifically, wire-cylinder and more specialized assemblies, such as those of the wire-duct and point-array plate type, are adopted to design electrostatic precipitators [see, for example, Adamiak (2013), Zheng et al. (2010), and Nouri et al. (2012)], ionizers [Intra and Tippayawong (2009)], and ozone generators [Shin and Sung (2005)]. This chapter is presented as a cursory survey of the available studies on the unipolar ion flow which, unfortunately, suffer from being tied to a restrictive number of canonical case studies. That is, the pair of wire-plane and point-to-plane coronas have perhaps been selected with the belief that they are the best informative, rather than representative, examples among the many coronas that could come to fruition. As will be appreciated in Chapter 5, the extended collection of setups explored there has been crucial for giving substantial arguments for a definitive revision of the ion-flow theory (Chapter 4).

2.2 THE UNIPOLAR SPACE-CHARGE FLOW PROBLEM

2.2.1 General Formulation

In brief, the field equations governing unipolar ion flows in the interelectrode space are expressed elsewhere in the form of simplified Maxwell's equations.

These legitimately neglect magnetic effects because of the slow-moving (thus, largely subsonic) flow of ions in the drift region. Accordingly,

$$\nabla \cdot \mathbf{E} = \frac{\rho}{\varepsilon_0} \qquad \text{Poisson's equation} \tag{2.1}$$

$$\mathbf{J} = k\rho\mathbf{E} \qquad \text{Ion current density} \tag{2.2}$$

$$\nabla \cdot \mathbf{J} = 0 \qquad \text{Current continuity(no ionization)} \tag{2.3}$$

For the sake of completeness, consider that even thermal diffusion is assumed to be unimportant, as suggested by Felici (1963) and Sigmond (1986). The unknowns of the problem are the vector fields $\mathbf{E}$ and $\mathbf{J}$ and the scalar field represented by the charge density ρ. The mobility k is expressed as a unique average quantity in view of the different ionic species really present in the drift region; ε_0 denotes ambient permittivity, here taken equal to that of free space. Introducing the electrostatic potential φ and combining the related definition $\mathbf{E} = -\nabla\varphi_E$ with Eqs. (2.1)–(2.3) gives the generally unamenable nonlinear third-order partial differential equation

$$\nabla \cdot \left(\nabla\varphi_E \nabla^2 \varphi_E \right) = 0 \tag{2.4}$$

applied to steady ion flows. As the electric field $\mathbf{E}$ is related to the charge density ρ (see Section 2.2.3), even solving Poisson's equation in conjunction with the continuity equation results in a very difficult exercise. Qualitatively speaking, the Poissonian field is expected to depart in strength and orientation from the Laplacian field, throughout, owing to the nonzero ρ-field also present in the domain under examination. It is tacitly understood that the source of the referential Laplacian field is the charge deposited in electrostatic equilibrium on the surface of the electrode system when a given voltage is applied to it. Being of the third order, Eq. (2.4) can in principle be resolved by imposing Dirichlet or Neumann boundary conditions supplemented by a distribution law, governing the electric field or one of the parameters representative of the ion injection, at the emitter. The above additional condition can be set by evoking Kaptzov's hypothesis for the electric field (Section 1.10) or, say, assigning the current distribution. Unfortunately, the latter can only be strongly arbitrary owing to the lack of knowledge of the injection mechanism, which in turn is subject to the shape of the ionization region. However, there is a very restricted number of cases, all displaying geometrical symmetries, in which approximate analytical solutions of Eq. (2.4) have been successfully found. Townsend (1914) was able to solve Eq. (2.4) for the special coaxial-cylinder configuration after introducing the low-current approximation; Felici (1963) studied a few cases making recourse to separable solutions applied to simple geometries (a pair of grounded parallel planes or a pair of concentric spheres) and proved the existence of an often mentioned saturated or asymptotic solution, namely, one obtained when some favorable physical circumstances safely lead us to neglect the influence of the background field. This component is due to the charges residing in electrostatic equilibrium on the electrode surfaces. Indeed, saturated (or almost saturated) conditions are even found under usual conditions, with the exception of an often restricted portion of the drift region in close proximity to the active conductor. On the other hand, when the applied voltage is only slightly exceeding the corona onset level, the resulting weak corona gives rise to a proportionally negligible space-charge distortion, throughout, so that the distribution of charge density is given by *a priori* imposing a Laplacian-field pattern. However, Canadas et al. (1975) showed that in any corona geometry an increasing unipolar ion current rapidly switches the field lines toward a saturated asymptotic configuration, the geometry of which is not radically different from the Laplacian one. Gary and Johnson (1990) introduced the notion of degree of corona saturation, which is an empirical measure of how close to saturated conditions a line subject to corona is operating. The space charge released by the emitting surface flows toward the inactive electrode as if it is convoyed into a number of joined fluxtubes collectively covering the electrode spacing. The higher the corona severity, the greater the charge quantity flowing to the ground through individual fluxtubes. The saturated condition is attained when the electric-field potential due to the space charge within a given fluxtube, calculated by field integration from the ground to the emitter along the median fluxline of the fluxtube, is equal to the applied voltage. In a purely ideal sense, no further charge can be released by the active conductor because the electric field at its surface remains equal to the onset value. Accordingly, corona current, space-charge density, and electric field assume values that cannot be surpassed, a circumstance favorable to establish a two-sided bound formulation for any generic corona activity whose bounds are the weak and saturated coronas. Indeed, the degree of corona saturation depends on the surface electrode's real conditions and does not depend on the electrode geometry or applied voltage. This explains the ease of reproduction and extrapolation of electrical quantities from small-scale models. The actual on-ground electric field and current density can be calculated by using empirical formulas available in the EPRI Report (1982) from those given under established saturation conditions.

2.2.2 Iterative Procedure

Atten (1974) developed a systematic iterative procedure for solving Eq. (2.4) for any geometrical configuration. By combining Eqs. (2.1)–(2.3), he obtained the following two equivalent equations:

$$\nabla^2 \varphi_E = \frac{-\rho}{\varepsilon_0} \qquad (2.5)$$

with boundary conditions $\varphi_e = V$ and $\varphi_g = 0$ on the high-voltage and grounded electrodes, respectively, and

$$\varepsilon_0 \nabla \varphi_E \cdot \nabla \left(\frac{1}{\rho} \right) = 1 \qquad (2.6)$$

The procedure consists in assuming for ρ a reasonable initial distribution $\rho = \rho_0$ and solving Eq. (2.5) for φ_E; then, this solution is inserted into Eq. (2.6) to find the next approximate value of ρ. This alternative way of dealing with Eq. (2.4) has been used by several authors for implementing numerical procedures, all aimed at solving the corona-sourced ion-flow problem.

2.2.3 The Unipolar Charge-Drift Formula

The idea of the charge-drift formula was initially found in the works of Waters, Rickard and Stark (1970) and, independently, Sigmond (1978). Using Maxwell's equations gives

$$\frac{\partial \rho}{\partial t} + \nabla \cdot \mathbf{J} = \frac{\partial \rho}{\partial t} + \nabla \cdot (\rho \mathbf{v}) = 0 \qquad (2.7)$$

which represents the charge-continuity equation relative to a fixed volume through which the charge moves. Here, t is the time and $\mathbf{v}$ is the velocity impressed to the charges. An alternative approach, where the volume is assumed to move while containing the same amount of charge, leads one to write

$$\frac{d\rho}{dt} = \frac{\partial \rho}{\partial t} + \mathbf{v} \cdot \nabla \rho = \rho \nabla \cdot \mathbf{v} \qquad (2.8)$$

where $\mathbf{v} = k\mathbf{E}$ and $\mathbf{J} = k\rho\mathbf{E}$. Accordingly, the charge-drift equation

$$\frac{d\rho}{dt} = -\frac{k}{\varepsilon_0} \rho^2 \qquad (2.9)$$

is obtained. Note that ρ obeys the time-dependent law

$$\frac{1}{\rho(t)} - \frac{1}{\rho(0)} = \frac{kt}{\varepsilon_0} \tag{2.10}$$

where t specifically represents the time of flight subsequent to the instant $t = 0$ in which $\rho = \rho(0)$. The validity of this formula depends directly on the fact that the mobility coefficient μ is assumed to be constant (bear in mind that k is given by averaging the mobilities k_i of different ionic species, each bearing a charge whose volume density in the interelectrode gap is ρ_i).

Equation (2.10) shows that the asymptotic distribution $\rho(t)$ is approximately ε_0/kt, which Sigmond (1986) termed the saturation unipolar ion density. Under such circumstances, charged clouds tend to a uniform density ε_0/kt as time t increases. Jones et al. (1990) used Eq. (2.10) to simulate the current density distribution on a grounded planar collector (see Appendix 2.A) by implementing an expanding sphere model (see Section 2.4).

2.3 DEUTSCH'S HYPOTHESES (DH)

Townsend (1914) rigorously analyzed the ionized field for unipolar problems in a simple configuration with concentric cylinders. Because of the given cylindrical symmetry, the analysis is reduced to a 1D problem. Deutsch (1933) extended the analysis to the parallel-plate (1D), concentric-spherical (1D), and wire-plane (2D) configurations. This investigator established the following three hypotheses, some of which are applicable under specialized conditions, regarding the unipolar ion flow in the wire-plane gap:

1. The space charges affect only the magnitude but not the direction of a referential space-charge-free (Laplacian) electric field.
2. The charge density remains constant along any field line.
3. The electric field at the passive electrode remains at its Laplacian value.

Assumptions 2 and 3 are presumed to be permissible with the stipulation that the applied voltage is slightly higher than the corona onset value, whereas the first one, extensively termed the Deutsch assumption without any further specifications, is employed in the simulation of generally configured ion-flow fields in the space. The attractive consequence is that an originally 3D generic problem is, as a matter of fact, reduced to a 1D one because it is methodically applied to individual Laplacian trajectories.

In this section, we only deal with Assumption 1, which, differently speaking, admits that the ion trajectories in the drift region are guided by the Laplacian pattern. Therefore,

$$\mathbf{E} = \ell \mathbf{E}_L \tag{2.11}$$

where ℓ is a scalar function of position and $\mathbf{E_L}$ is the Laplacian field. By substituting Eq. (2.11) in Eqs. (2.1)–(2.3) we obtain

$$\nabla \cdot \mathbf{E} = \mathbf{E}_L \cdot \nabla \ell = \mathbf{E}_L \frac{\partial}{\partial s} \ell = \frac{\rho}{\varepsilon_0} \tag{2.12}$$

and

$$\nabla \cdot \mathbf{J} = \nabla \cdot (k\rho\ell\mathbf{E}_L) = \mathbf{E}_L \cdot \nabla(k\rho\ell) = \mathbf{E}_L \frac{\partial}{\partial s}(k\rho\ell) \tag{2.13}$$

Since $k = \mathrm{const}$, then Eq. (2.13) gives $(k\rho\ell) = J/\mathbf{E}_L = \mathrm{const}$ along any given fieldline (here otherwise representing an ion-drift trajectory). Moreover,

$$\nabla \times \mathbf{E} = \mathbf{E}_L \times \nabla \ell = 0 \tag{2.14}$$

which implies a parallelism between $\nabla \ell$ and the Laplacian field. In other words, the function $\ell = \ell(\varphi_L) = \mathrm{const}$ along each Laplacian equipotential surface.

After Usynin (1966), combining Eqs. (2.12) and (2.2) gives

$$\rho = \varepsilon_0 \mathbf{E}_L \frac{\partial}{\partial s} \ell = -\varepsilon_0 \mathbf{E}_L^{\,2} \frac{d}{d\varphi_L} \ell \tag{2.15}$$

$$J = k\rho\ell\mathbf{E}_L = -k\varepsilon_0\ell\mathbf{E}_L^{\,3} \frac{d}{d\varphi_L} \ell \tag{2.16}$$

which are functions of ℓ and $d\ell/d\varphi_L$. It is worth mentioning that some authors, favorably disposed toward DH, have solved the above equations by implementing the usual numerical procedures such as the Finite Element Method and the Finite Difference Method.

2.4   SOME UNIPOLAR ION-FLOW FIELD PROBLEMS

Special reference will be made here to the wire-plane and point-to-plane assemblies for the reasons previously explained in Section 2.1. Figures 2.1

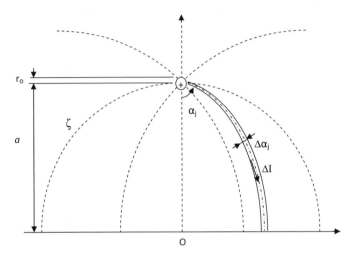

FIGURE 2.1 Unipolar wire-plane configuration. The injecting angle α_j selects the dashed median fluxline of an elemental fluxtube (full line) which carries a partial quantity ΔI, proportional to the angular width $\Delta \alpha_j$, of the total current I.

and 2.2 show the main geometrical and electrical quantities for both configurations (see captions to the figures for details). Several authors based their approach on Deutsch's assumption for obtaining approximate analytical [see Popkov (1953, 1980, 1981). Canadas et al. (1975), Tsyrlin (1957, 1958), Usynin (1966), Sigmond (1982), Walsh, Pietrowski, and Sigmond (1984)] or numerical [Atten (1974), Sarma Maruvada and Janischewskyj (1969), Yang,

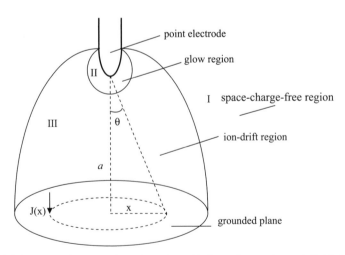

FIGURE 2.2 Point-plane configuration. Overall electrode spacing subdivided into three compartments. [Reproduced from Jones et al. (1990) with permission from *Journal of Physics D: Applied Physics*.]

Lu, and Lei (2008)] solutions. In both cases, the very purpose of the above investigations was to determine some performances, notably the voltage–current curve and the pair of current density $J(x)$ and electric-field $E(x)$ distributions on the earthed plane. Other authors have discussed the validity of DH and at the same time tried to implement some procedures for solving the unipolar space-charge flow problem after waiving DH.

2.4.1 Analytical Methods

As stated previously, the Poissonian and Laplacian fluxlines become super-imposed, in agreement with DH. In the wire-plane case, the suspended conductor is assumed to be a perfectly horizontal cylinder. Provided that the wire elevation a is far exceeding the conductor radius ($a \gg r_0$; see Figure 2.1), a bifocal cylindrical coordinate system can be profitably applied to an imaged two-parallel, indefinitely long straightline model. Accordingly, the Laplacian fieldlines of the original system approach arcs of circumference with their centers lying on the ground plane at a distance from the coordinate system origin 0 equal to ζa (the dimensionless quantity ζ could label a generic fieldline; see again Figure 2.1). The orthogonal equipotential lines are circumferences whose centers are located on the vertical line through 0 at height $a(\zeta^2 + 1)/(\zeta^2 - 1)$.

In the point-to-plane (also termed rod-plane) configuration, a typical approximation consists in assuming that the shape of the active electrode fits a hyperboloid. By using a prolate-spheroidal reference frame, the Laplacian field can be described by two orthogonal families of curves. These represent fieldlines and equipotential lines respectively forming a pair of confocal ellipsoid (focal length h_f) and confocal hyperboloid sets. Introducing the parameter $\omega = \text{const}$ for each equipotential line of the family gives the gap separation $a = h_f \cos\omega$, thus nearly equal to h_f when the electrode is very thin ($\omega \to 0$). Details of the graphical field construction can be found in Durand (1964), Moon and Spencer (1961), and Allibone et al. (1993).

As previously mentioned, additional quantities of interest could be the planar profiles of current density, electric field, and charge density. The analytical methods used by several authors to give resolving closed forms are reported in Sigmond (1986) in conjunction with a list of approximate formulas (here reproduced, for the reader's convenience, in Tables 2.1 and 2.2) for the current density at the ground plane in the wire-to-plane and point-to-plane configurations. In particular, even the total current I_T lost at the wire for wire-plane coronas is given. The celebrated Townsend's (approximate) solution for a coaxial geometry reads $I_T = F_{\text{cyl}}V(V - V_T)$, where V is the applied voltage, V_T the onset voltage, and F_{cyl} a parameter dependent on the system geometry and ion mobility. The validity of this typically quadratic law, customarily

TABLE 2.1   Planar current density (approximate formulas)

	Wire-plane	Notes
Deutsch	$J(\theta) = \dfrac{k\varepsilon_0}{\lambda_d\, a^3} V(V - V_T)F_D(\theta)$	D curves in Figure 2.3
	where	Corona current/m:
	$F_D(\theta) = (1 + \cos 2\theta) \displaystyle\int_0^{\lambda_d} \dfrac{\gamma\, d\gamma}{(\cos\gamma + \cos 2\theta)^2}$	$I_D = 9.8\dfrac{k\varepsilon_0}{\lambda_d a^2} V(V - V_T)$
	and $\lambda_d = \ln\left(\dfrac{2a}{r_0}\right)$	
Popkov (L)	$J(\theta) = \dfrac{k\varepsilon_0}{2\lambda_d^{\,2} a^3}(V_T)^2 M_b$	PL1 and PL2 curves in Figure 2.3
	where M_b is a dimensionless quantity obtained by numerical integration of the following equation:	
	$\dfrac{V}{V_T} = \displaystyle\int_0^1 \sqrt{M_b \int_u^1 1 + \dfrac{du'}{e_L^2}}\, du$	
	where $\displaystyle\int_u^1 \dfrac{du}{e_L^2} = \dfrac{1}{(1 + G)(1 - G^2)}\cdot$	
	$\cdot\left[1 - \dfrac{\sqrt{x^2 - 1}}{x + G} + \dfrac{G}{\sqrt{1 - G^2}}\right.$	
	$\left.\left(\dfrac{\pi}{2} - 2\theta - \arcsin\dfrac{1 + Gx}{x + G}\right)\right]$	
	and $e_L = E_L/V; u = \varphi_L/V; G = \cos 2\theta$	
Popkov (T)	$J(\theta = 0)\dfrac{(1 + \cos^2\theta)\sin^3(2\theta)}{6(\sin 2\theta - 2\theta\cos 2\theta)}$	PT curve in Figure 2.3
Sigmond	$J(\theta = 0)\dfrac{\sin(2\theta)}{2\theta}$	S curve in Figure 2.3
		Corona current (saturation):
		$I_S = 1.62\dfrac{k\varepsilon_0}{a^2} V^2$
Usynin/Walsh –Gallo–Lama	$J(\theta = 0)\cos^6\theta$	U curve in Figure 2.3

TABLE 2.2   Planar current density (approximate formulas)

	Point-to-Plane	Notes
Deutsch	$$J(\theta) = \frac{k\varepsilon_0 V(V - V_T)}{a^3 k_a \ln 2 \cdot \xi_\theta \left(\xi_\theta^2 - \frac{1}{3} - \frac{1}{6} \ln 2 \right)}$$ where $$\xi_\theta = \frac{1}{\cos\theta} \quad \text{and} \quad k_a = \ln\left(\frac{1 + \eta_a}{1 - \eta_a} \right) \text{ with}$$ η_a being the spheroidal coordinate labeling the surface of the active tip	D curve in Figure 2.4; corona current $$I_D = \frac{11.85 k\varepsilon_0}{k_a a} V(V - V_T)$$
Popkov (L)	$$J(\theta) = \frac{k\varepsilon_0}{k_a^2 a^3} (V_T)^2 M_b$$ where M_b is a dimensionless quantity obtained by numerical integration of the following equation: $$\frac{V}{V_T} = \int_0^1 \sqrt{1 + M_b \xi_\theta \left[\xi_\theta^2 (1 - \eta') - \frac{1 - \eta'^3}{3} \right]}\, du$$ with $\eta' = \tan h \dfrac{k_a u}{2}$;	PL1 and PL2 curves in Figure 2.4
Popkov (T)	$$J(\theta = 0) \frac{2\cos^3\theta}{3 - \cos^2\theta}$$	PT curve in Figure 2.4
Sigmond	$$\approx J(\theta = 0) \frac{k\varepsilon_0 V_T^2}{\left(a\sqrt{1 + 2\tan^2\theta} \right)^3}$$	S curve in Figure 2.4; corona current (saturation): $I_S \approx \dfrac{2k\varepsilon_0}{a} V^2$
Walsh–Pietrowski	$$\approx J(\theta = 0) \frac{1 + 2(\xi_\theta - 1)/k_a}{\xi_\theta^5}$$	WP curve in Figure 2.4; Total current: $I_{WP} \approx \dfrac{2\pi}{3} \left(1 + 1/k_a \right) \dfrac{V^2}{a}$
Walsh–Gallo–Lama	$J(\theta = 0)\cos^3\theta$	WGL curve in Figure 2.4
Warburg	$J(\theta = 0)\cos^m\theta$	$m = 4.82$ positive corona $m = 4.65$ negative corona

called the I–V characteristic, was later confirmed by other authors and extended to other geometrical configurations, so Townsend's formula turns out to be a reference for generic numerical applications.

Current density, charge density, and ion-flow velocity are functions of spatial coordinates and, in particular, of the electric field. When the unipolar corona problem is solved and the electric field in the presence of space charge is known, all the previous quantities can be calculated. Therefore, an approximate evaluation of the above quantities under saturated conditions can be carried out by assuming, after Sigmond (1982), the average value of the electric field along any given ionic flight path of length L. In this way, the ionic velocity v, time of flight τ, charge density ρ, and current distribution J can be calculated at the plane by the following formulas:

$$v = \frac{kV}{L}, \qquad \tau = \frac{L^2}{kV}, \qquad \rho = \frac{\varepsilon V}{L^2}, \qquad J = \frac{\varepsilon k V^2}{L^3} \qquad (2.17)$$

Because the ion trajectory is a quarter of an ellipse for the hyperboloid-plane geometry, then by approximately imposing L coincident with the chord of the above elliptic arc, we derive

$$J(\theta) \approx J(\theta = 0)(1 + 2\tan^2\theta)^{-3/2} \qquad (2.18)$$

for the current–density distribution (see Table 2.2). The curves of the normalized formulas appearing in Tables 2.1 and 2.2 are traced in Figures 2.3 and 2.4 for a comparison with experimental data. In particular, Figure 2.4 shows the Warburg cosine law (WL) for the hyperboloid-to-plane configuration (see Appendix 2.A for a detailed description of this semiempirical law).

In the wire-to-plane geometry, the better agreement with experimental data is obtained by the formulas of Deutsch (coincident with Popkov L1) and Popkov (T) (see caption to Figure 2.3). In this figure, also the numerical results waiving DH, obtained by Janischewskyj, Sarma Maruvada, and Gela (1982), and Takuma, Ikeda, and Kowamoto (1981), are reported (see Section 2.4.2). In the hyperboloid-plane geometry, the better agreement with experimental data is obtained by using WL, Sigmond (saturated), and Popkov-L formulas. The proposed closed forms may be classified as in "longitudinal" (L) and "transversal" (T) types. The L-type formulas (Popkov-L, Deutsch, Sigmond, Walsh–Pietrowskj–Sigmond, Tsyrlin) assume that the current flows are guided by the Laplacian pattern. The consequence is that current continuity is preserved whereas the equipotential surfaces cannot become orthogonal to the fieldlines. Instead, the T-type formulas (Popkov-T, Usynin, Walsh–Gallo–Lama) assume that Poissonian equipotential surfaces are orthogonal to the field lines. The consequence now is that current continuity is violated.

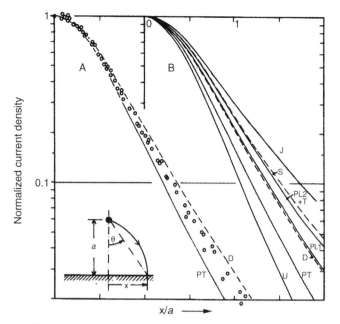

FIGURE 2.3   Wire-plane configuration. Normalized current density distributions as listed in Table 2.1. $\bigcirc$, Experimental data from Popkov and Ryabaya (1974); D, Deutsch approximation; PL1, Popkov L formula with $V/V_T = 1.05$ and $M_{b0} = 6$; PL2, Popkov L formula with $V/V_T = 4$ and $M_{b0} = 1900$; U, Usynin/Walsh–Gallo–Lama formula; PT, Popkov T formula; S, Sigmond (saturated) formula; J, Janischewskyj et al. (1982) by FEM; T, Takuma et al. (1981) by FEM. [Reproduced from Sigmond (1986) with permission from *Journal of Electrostatics*.]

Sigmond (1986) concludes in his review paper that the L-type solutions have to be preferred because "it seems less important to ensure that the equipotential surfaces stay normal to the Laplacian field lines."

Simplified expressions for ionized electric fields and corona losses pertaining to HVDC monopolar lines are also available in D'Amore, Daniele, and Ghione (1986). The proposed solution is based on DH and an integration procedure involving a second-order differential equation [obtained by combining Eqs. (2.1)–(2.3)]. Therefore,

$$\nabla \cdot (\mathbf{E} \nabla \cdot \mathbf{E}) = 0 \qquad (2.19)$$

is obtained [it is profitable to compare the above formula with Eq. (2.4)]. The charge and current densities at the emitter are expanded in a truncated Fourier series with two or three harmonics. Simple analytical formulas are furnished for practical lines, provided that the height of the conductors is very large with respect to the radius.

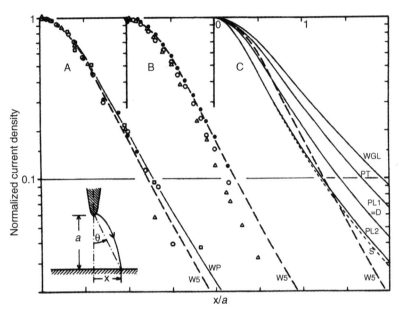

FIGURE 2.4 Point (hyperboloid)-plane configuration. Normalized current density distributions as listed in Table 2.2. *Points A*: measured data from Popkov, *Izv. Akad. Nauk USSR* 1980, **2**:95; ○ hyperboloid endpoint angle = 2.5° and $a = 4$ cm; ● and □, hyperboloid endpoint angle = 2.5° and $a = 20$ cm; △, hyperboloid endpoint angle = 8.5° and $a = 4$ cm. *Points B*: Goldman et al. (1978) measured data: ○, positive corona; ●, negative corona; $a = 12$ cm and $V = 40$ kV. Kondo and Miyoshi (1978) measured data: △, Negative pulseless glow, $a = 1.2$–1.4 cm and $V = 22$ kV. *Points C*: D, Deutsch approximation (coincident with PL1); PL1, Popkov L-formula with $V/V_T = 1.05$ and $M_{b0} = 14.5$; PL2, Popkov L-formula with $V/V_T = 8$ and $M_{b0} = 2200$; WGL, Walsh–Gallo–Lama formula; PT, Popkov T-formula; S, Sigmond (saturated) formula; Walsh–Pietrowski formula (coincident with WL $m = 4.65$); W5, Warburg law with $m = 5$. [Reproduced from Sigmond (1986) with permission from *Journal of Electrostatics*.]

Other researchers attempted to solve the unipolar ionized field problem without using simplifying hypotheses. Ciric and Kuffel (1982) used the bipolar coordinate system for solving Eqs. (2.1)–(2.2), in conjunction with the integral form of the continuity equation, in the wire-plane case study. At first, they adopted both DH and Kaptzov's hypothesis (KH; see Section 1.11), imposed $k = $ const, and neglected diffusion. Their analysis leads to express the electric field, charge density, and current density at any point of the drift region by closed forms, with special attention paid to those applicable at the ground level. The relation between V/V_T and I is expressed under integral form. Then the authors give several simplified I–V relations with differently restricted validity ranges and deduce Popkov's formula as a particular case of their equation. Later, the same authors improved their approach by retaining DH and eliminating KH in order to analyze the influence of various boundary conditions on the distributions of the electric field and current density [see Ciric and Kuffel (1983)]. They obtained a general integral form of the voltage–current relation for any specified elementary fluxtube. Furthermore, they considered three special types of boundary conditions at the surface of the conductor in the corona: (a) The electric field E_c is a known function of the surface points [if the electric field is constant, the same expressions shown in Sarma and Janischewskyj (1969) are obtained]; (b) the charge density ρ_c is a known function of the surface position, as in Khalifa and Abdel-Salam (1973) and Takuma, Ikeda, and Kowamoto (1981); or (c) the current density j_c is a known function of the surface position. In all the described cases, the given on-ground profiles are comparable to those previously published by other authors. In conclusion, the conditions adopted at the surface of the conductor in the corona happen to have a prominent influence on the ionized-field quantities. This approach also allows the evaluation of the boundary conditions from available measurements at ground level (inverse problem).

Aboelsaad, Shafai, and Rashwan (1989a,b) improved this approach by expressing the ratio E_c/E_0 on the surface points of the emitter by a dimensionless function f_1 that is obtained by using a second-order polynomial regression of the variation of average values. These are made available by measuring the electric field on the conductor surface [see Waters, Rickard, and Stark (1972), Popkov, Bodganova, and Pevchev (1978), and Abdel-Salam, Farghally, and Abdel-Sattar (1983)]. Figure 2.5 shows the corona I–V characteristics obtained by assuming that the surface electric field E_c changes on the conductor (curve a) and that $E_c = E_T$, irrespective of the applied voltage (curve b). Forcing KH in the analysis causes an underestimation of corona losses. Li and Wintle (1989) used a conformal transformation for solving Eqs. (2.5)–(2.6) in combination with a finite difference procedure. The purpose was to map the domain under examination for the wire-plane configuration. They avoided both DH and KH, otherwise assuming an assigned distribution of charge density $\rho = \rho(r, \theta)$ taken elsewhere. For the general definition of the semi-vertical cone or wedge angle of discharge θ, use could be made of Figure 5.A.1 in Appendix 5.A (Chapter 5); see also Figures 2.1 and 2.3 for the special wire-plane case, where θ specializes in a semi-vertical wedge angle of discharge, and Figures 2.2 and 2.4 for the special point-plane case, where θ specializes in a semi-vertical cone angle of discharge. They were able to draw a number of issues, among which it could be stressed that the electric field on the surface of the emitting wire is rather a function of the angle θ and much lower than that given by using Peek's formula; the contours of the charge density surrounding the wire change from an ovoidal (at low densities and at some distance from the wire) to a cardioid shape (at high densities and in close proximity to the wire); the electric field at the collector is proportional

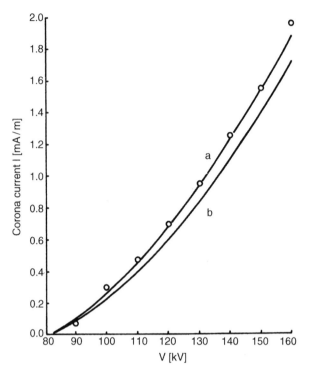

FIGURE 2.5 Cylindrical configuration. Corona I/V characteristics [see Aboelsaad et al. (1989a), with permission from IEE]. Curve a: $E_c = f_1 E_T$. Curve b: $E_c = E_T$ (according to KH); ○ experimental data from Waters et al. (1972).

to $\cos^2\theta$, whereas the current density at the collector is proportional to $\cos^4\theta$. Apart from the exponents involved, these laws do evoke Warburg's empirical cosine law, whose generalization of the kind $\cos^n\theta$, with n changing in each individual example, was restrictively considered applicable to the point-plane configuration alone, in which case n nearly approaches 5. Indeed, as will be appreciated in Chapter 5, the above restriction appears debatable. Last, the authors are inclined to conclude that the distributions at the ground plane are rather similar in all the cases, thus slightly dependent upon the boundary conditions established at the distanced wire.

2.4.2 Numerical Methods

Different numerical procedures have been proposed in the literature to show the validity of some simplifying hypotheses and the ways to abstain from using them for solving unipolar flow problems. These procedures are summarized as follows.

(a) DH- and KH-Based Approaches   Sarma Maruvada and Janischew-skyj (1969) developed a numerical method, called the Flux Tracing Method (FTM), substantially based on DH and KH. Later, Sarma Maruvada (2012) used the FTM and compared the given numerical results with available experimental data from a DC test line and three operating DC transmission lines. The paper only considers calculation methods for unipolar-space-charge-modified fields. Further simplifying hypotheses are required for reducing the computational efforts in simulating the transmission line as a source of bipolar space charge. In practice,

- The transmission-line configuration is represented by infinitely long cylindrical conductors, all parallel with the ground plane.
- The drift region fully occupies the space between the conductors and ground plane (i.e., the thickness of the corona ionization layer is assumed negligible).
- Ionic mobility is constant and diffusional effects are assumed unimportant.
- The effects of any exogenous wind and ambient parameters (humidity, aerosols, etc.) are neglected.

Consequently, the field is described by Eqs. (2.1)–(2.3), and, according to DH, the equation system to be solved is formed by the following differential equations, as deduced from Eqs. (2.15) and (2.16):

$$\frac{d\varphi_E}{d\varphi_L} = \ell \tag{2.20}$$

$$\frac{d}{d\varphi_L}\ell = -\frac{\rho}{\varepsilon_0 E_L^2} \tag{2.21}$$

$$\ell\frac{d\rho}{d\varphi_L} = \frac{\rho^2}{\varepsilon_0 E_L^2} \tag{2.22}$$

The boundary conditions are expressed by

$$\varphi = \varphi_L = V_c \qquad \text{at the HV conductor} \tag{2.23}$$

$$\varphi = \varphi_L = 0 \qquad \text{at the grounded conductor} \tag{2.24}$$

$$\ell = \frac{E_T}{E_{c,L}} \qquad \text{at the HV conductor} \tag{2.25}$$

where E_T is the corona onset value at this electrode and $E_{c,L}$ is the (possibly changing) space-charge-free electric field at the surface of the same electrode raised at the potential V_c.

The main steps adopted to implement the calculation of the electric field and current density are listed in succession as follows:

1. For the DC line configuration and specified conductor voltages, a reliable method of successive images is applied in Sarma Maruvada (2000) to determine the space-charge-free electric field.

2. For a given point on the ground plane, the fluxline of the space-charge-free electric field is traced and the point of contact of the fluxline on the conductor is then identified.

3. For each traced fluxline, normalized values of E_L and φ_L (for equal steps of $\Delta\varphi_L$) are calculated, starting from the conductor surface, and then stored.

4. Equations (2.20)–(2.22) are solved iteratively to satisfy the boundary conditions, expressed by Eqs. (2.23)–(2.25), taking into account an initial estimate of ρ at the conductor surface. Along each fluxline, the values of E and ρ in function of φ_L are obtained.

5. The current density $J = k\rho E$ can be calculated throughout the fluxline.

By an iterative procedure involving steps 2–5, the complete profile of $J(x)$ on the grounded conductor is furnished.

Because the adopted FTM takes into account the debated DH, a check is introduced for estimating the error affecting the equipotential pattern. Once the potential φ_E has been given by solving Eqs. (2.20)–(2.22) and meeting the conditions imposed by the form of Eqs. (2.23)–(2.25), its pattern has been compared with that pertaining to the space-charge-free case to evaluate possible departure. This has been assumed as an indicator of DH failure in simulating a bipolar DC test line (see Appendix 2.B). It has been verified that the error introduced by DH is prone to increase in proximity to the ground plane, perhaps where the current density and the electric field profiles turn out to be of major interest. Notwithstanding this, further checks on operating bipolar DC transmission lines and comparison with available experimental data definitively led the author to state the following:

- The distributions of the ground-level electric field and ion current density given by a DH-based FTM are reasonably accurate, particularly in the unipolar regions just below the conductors where profile maxima are positioned.

- For practical DC transmission lines, calculations performed by using the FTM show that the roughness factor m assigned to the surface of the active conductor significantly influences the distributions on the inactive planar collector.
- The errors imputable to DH may be considered negligible if compared to other causes of uncertainty affecting computational prediction and measurement. As a consequence, what has previously been referred to as DH could in practice appear to be a problem that is not genuine.

(b) Approaches Waiving DH

(B1) Finite Element Method (FEM)   Perhaps the first attempt to solve the problem of calculating the unipolar-space-charge-modified field without retaining DH is found in Janischewskyj, Sarma Maruvada, and Gela (1982). For this purpose, a FEM subject to KH as a boundary condition for the electric field on the emitting surface was employed. Their analysis offered arguments for discussing how DH influences the evaluation of corona losses and on-plane electric field distributions. For the case of a unipolar DC corona, the set of Eqs. (2.1)–(2.3) is given in terms of potential φ, as follows:

$$-\nabla \cdot \nabla \varphi_E = \frac{\rho}{\varepsilon_0} \tag{2.26}$$

$$-\nabla \cdot (\rho \nabla \varphi_E) = -[\nabla \rho \nabla \varphi_E + \rho \nabla \nabla \varphi_E] = 0 \tag{2.27}$$

$$\nabla \varphi_E = -\mathbf{E} \tag{2.28}$$

This new set of equations is solved iteratively by the FEM for φ_E, and the spatial distribution of charge ρ is recomputed at the beginning of each iteration. The initial distribution of charge is established through the approach in Sarma Maruvada and Janischewskyj (1969) based on DH. The convergence of the process is achieved when both no further improvements are observed for ρ and Eqs. (2.27) and (2.28) give the same values of φ_E. However, there are two sources of error in this FEM-based algorithm: (a) The domain of analysis is bounded by an artificial surface surrounding the DC line configuration; and (b) as the solution is obtained in terms of potential φ, the boundary condition expressed in terms of electric field requires a numerical differentiation of φ. Moreover, as unenforced during the iterative procedure where Eqs. (2.26) and (2.27) are involved, the given boundary condition may be violated in any iteration. The extent of this violation provides suitable information for correcting the distribution of ρ for the next iteration, until convergence is accomplished. The source of error (b) above has been removed in Qin, Sheng, and Yan (1988) by an improved approach (which will be described later on) in which the solution is directly expressed in terms of **E** instead of electric field and current profiles at the planar collector, as found in Janischewskyj, Sarma Maruvada, and Gela (1982). At first, the authors simulate the same configuration adopted by Popkov (1949) in his experimental studies and conclude that the electric field profile at ground is rather insensitive to DH, even after increasing the applied voltage. Conversely, the current density profile appears significantly influenced by the above hypothesis in the sense that the more voltage is increased, the larger is the difference between the two planar current profiles obtained by using and waiving DH. In any case, the profile based on DH is corroborated by the measured data in Popkov (1949).

Further comparisons between simulation results (obtained without adoption of DH) and data supplied by tests on a unipolar line and an operating bipolar DC transmission line are also available in Janischewskyj, Sarma Maruvada, and Gela (1982). The authors conclude that DH may cause more errors in solving unipolar than bipolar ionized fields can do, because in this case positive and negative ions mix and recombine as to reduce the density of the net space charge. Therefore, a comparison with experimental data gives additional arguments in favor of the practical validity of DH in designing HVDC transmission lines.

Also other investigators [see again, for example, Takuma, Ikeda, and Kowamoto (1981)] used the FEM to solve the space-charge-modified problem described by Eq. (2.26) in combination with Eq. (2.27). In this case, DH has been waived and KH replaced with a boundary condition expressed in terms of constant ion density distribution at the emitter. In this regard, use has importantly been made of the equation

$$\nabla \rho \cdot \nabla \varphi_E = \frac{\rho^2}{\varepsilon_0} \qquad (2.29)$$

The numerical instability consequent to differentiation has been proved to be prevented. The case studies regard coaxial cylinders, unipolar and bipolar HVDC lines in the presence/absence of wind. The results relative to the unipolar wire-plane case, reported in Figure 2.3, permit appreciation of a current density profile in good agreement, thus better than what Janischewskyj's solution can do, with the PL2 curve and with experimental data. Overall, assuming a constant charge density on the emitter surface turned out to be a recommendable practice whenever the emitter is substantially distanced from the collector. Either way, this is not the case when the emitter is in the form of a bundle. Indeed, there are now substantial arguments (see Section 4.6) for informing the reader in advance that the above assumption is questionable.

Other solutions using finite-element algorithms have been proposed [see Aboelsaad, Shafai, and Rashwan (1989a,b) and Abdel-Salam and Al-Hamouz (1994)]. In the former reference, Eqs. (2.26) and (2.27) are simultaneously solved by introducing a polynomial formulation of the KH derived from previous numerical and experimental studies, such as those made by Waters, Rickard, and Stark (1972); in the latter, only the second-order PDE (2.26) is solved with the stipulation of respecting current continuity. In particular, the distribution of the charge density ρ_i in the interelectrode space is calculated step-by-step starting from the approximate on-plane distribution of ρ_e supplied elsewhere [Sarma Maruvada and Janischewskyj (1969)] and then proceeding toward the source. Pinning down the electric field on the emitting surface at a fixed value equal to the onset value E_T (KH) has been made feasible in the FE formulation. Hence, the procedure calculates the field (and the potential) at the successive nodes in each fluxtube by a third-order interpolating polynomial. The last estimates of the potential in close proximity to the ground are compared, and the average value of the nodal potential gives the relative error. This allows the space-charge density ρ_e to be corrected if the error exceeds an assigned value. Then, the space-charge density at different nodes along each fluxtube is corrected in view of current continuity. The procedure is of course reiterated until the error becomes smaller than the above-mentioned value.

The FEM has been implemented by Jones and Davies (1992) also for solving Poisson's equation when the point-to-plane geometry is taken into account. The entire domain under examination is partitioned into three regions where the following distinctive features can be appreciated (see Figure 2.2):

- Region I could be thought as a corona-free zone. Theoretically, the charge is in reality present in the whole space around the electrode in the corona, but the charge quantity could be assumed to be negligible in Region I. This legitimizes the application there of Laplace's equation $\nabla^2 \varphi_L = 0$ for the field mapping. The boundary conditions applicable to this region are $\varphi_L = 0$ on the grounded plane, $\varphi_L \rightarrow 0$ at large distances, $\varphi_L = V_T$ on the point electrode, and $\partial \varphi_L / \partial n = 0$ on the boundary fieldlines delimiting Regions I and III. As the presence of space charge in the gap (notably, in Region III) causes distortion of the Laplacian fieldlines, a reliable identification of such a contour poses difficulties because this can only be made through Poisson's equation. Otherwise, the recommended practice consists of using the Laplacian pattern smartly.

- In Region II, a complex ionization mechanism takes place (see a detailed description in Chapter 1). This activity is confined to a small region around the point where the electric field is high enough to make the difference $(\alpha - \eta)$ positive and the discharge a self-sustained one. Even though the thickness and contour of the ionized space cannot be exactly defined, presumable miscalculations are passed over owing to the restricted dimensions of that region compared to those labeled I and III. Owing to a balanced presence of positive ions and electrons, the net charge tends to vanish where ionization is being developed so that the resolving equation is reduced to the Laplacian one, subject to the condition $\nabla \varphi_L = E_c$ (according to KH, the electric field holds equal to the onset value E_c in each point of Region II).

- The charges injected from Region II into Region III drift, obeying Poisson's and current continuity equations. Accordingly,

$$\nabla \cdot \mathbf{J} = \nabla \cdot (k\rho\mathbf{E}) = -k\nabla \cdot (\rho\nabla\varphi_L) = 0 \qquad (2.30)$$

The boundary conditions are $\varphi_L = 0$ on the grounded plane, $\partial\varphi_L/\partial n = 0$ on the fieldline separating Regions I and II, and $\partial\varphi_L/\partial n = E_c$ on the ionization edge.

The electric field $\mathbf{E}(x)$ and the current distribution $J(x)$ on the plane are involved in the relationship $J(x) = k\rho(x)E(x)$, where a constant mobility k of known value is tacitly understood and $\rho(x)$ is the local charge density. Incidentally, some measurements of the electric field $\mathbf{E}(x)$ made by Selim and Waters (1980) lead to a good theoretical fitting under the form of the WL-like cosine law of the type $\mathbf{E}(x) = \mathbf{E}(0) \cos^p\theta$ with index $p = 2$. The numerical procedure in the paper under examination starts with assigning a rough finite-element Laplacian-field solution $\mathbf{E}_L(x)$. As k is given, the ratio $J(x)/(k\,\mathbf{E}_L(x))$ can be used as a boundary condition for $\rho(x)$ in a thin charge-layer joining the plane. A charge-layering method is then applied to proceed toward the electrode in the corona by using the Laplacian pattern for the nodal arrangement. After giving a first approximated evaluation of the space-charge density, the numerical procedure recalculates the electric field relative to this space-charge distribution and gives a new space distribution of nodes. The method goes on in an alternating fashion, giving increasingly refined distributions of the space charge and, in turn, of the electric field. When the convergence to the Poissonian distribution is reached with an assigned accuracy, the final nodal geometry can be drawn. Using WL with exponent $m = 4.65$ as a referential current distribution for negative coronas, the investigators found that the best fittings for $\mathbf{E}(x)$ and $\rho(x)$ can be obtained when WL-like formulas are adopted, with the specification that 1.8 and 2.85 are the exponents of the cosine laws that realistically reproduce, respectively, the experimental electric-field and charge density distributions at the planar collector.

(B2) Charge Simulation Method (CSM) Horenstein (1984) used the CSM to compute both space-charge-free and space-charge-modified electric fields under 2D circumstances. The thickness of the ionization layer is neglected in comparison to the gap extension, and the KH is accommodated. A space-charge-free solution satisfying KH at the surface of the active electrode is at first found. This solution is obtained by positioning charges inside the electrode (Laplacian solution). Then, for a given level of corona current, equipotential charge shells, represented by discrete line charges, are added to the original volume so as to enclose the electrode (see Figure 2.6). A vanishing contribution to the surface field, according to KH, is imposed for specifying these equipotential charge shells. The discrete space charges q_m have to be chosen in order for the potential

$$\Phi(\bar{r}_n) = \sum_{m=1}^{M} \frac{q_m}{2\pi\varepsilon_0} \ln \frac{1}{\bar{r}_n - \bar{r}_m} \tag{2.31}$$

calculated along each equipotential line to be constant. Here, $\bar{r}_n$ and $\bar{r}_m$ label locations for a selection of M points spread over the equipotential contour and

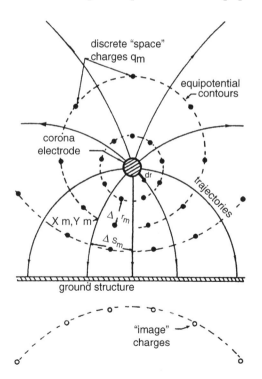

FIGURE 2.6 Wire-plane configuration. Application of the CSM and schematic representation of discrete space charges, fieldlines, and equipotential contours. [Reproduced from Horenstein (1984) with permission from IEEE.]

line-charges, respectively. Also, the discrete space charges q_n have to be chosen in order for the discretized form of continuity equation

$$\sum_{m=1}^{M} \frac{q_m}{\Delta r_m \Delta s_m} kE\Delta s_m \cong I_c \qquad (2.32)$$

to be satisfied. In this case, Δr_m is the distance between adjacent shells, Δs_m being the chord length between adjacent charges in the same shell and I_c the corona current per unit length. The results, given in terms of a I–V characteristic, are compared with the analytical solution relative to a coaxial-cylinder configuration. An increase of the shell number from 100 to 1000 leads to a reduction of the error, referred to an analytical solution, from 0.3% to 0.03%. However, a good agreement with experimental data was obtained by only using 10 equipotential shells for different values of the corona current. The small number of shells involved, in combination with the fact that iteration techniques are circumvented, makes this CSM-based method especially recommended as to the short computational time required.

Qin, Sheng, and Yan (1988) applied the CSM to a bipolar configuration (see Figure 2.7). The involved Eqs. (2.1)–(2.3) are estimated as being suitable for positive and negative regions, but in need of improvement for the bipolar region (see Appendix 2.B, where the wind action and diffusion for positive and negative ionic species are also taken into account). To avoid the error due to the numerical differentiation of φ introduced by the FEM, the CSM is used to compute the component of **E** directly from Eq. (2.31). Bearing in mind that the total electric field **E** is composed of the space-charge-free electrostatic field $\mathbf{E}_L$ and the field $\mathbf{E}_{ch}$ is produced by the corona-generated space charge, then $\mathbf{E}_L$

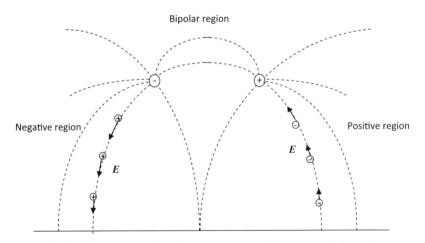

FIGURE 2.7 Bipolar wire-plane configuration. Fluxlines and charge flow.

is computed by replacing the conductors with simulating charges located within the conductors. The values of these charges must satisfy the Dirichlet boundary conditions on the conductor surfaces and the zero-potential condition on the ground. The further quantity E_{ch} may easily be computed by using Gauss' law, in which case the space charges ρ_+ and ρ_- are represented by discrete values at the node of a grid preliminary superimposed to the domain under examination. The given electric field E is involved in equations, solved through the method of weighted residuals, for ultimately yielding fresh distributions of ρ_+ and ρ_- which in turn modify $E(P)$ in any point P, and so on. As usual, the iteration stops when two subsequent sets of charge distributions differ of a pre-established small quantity. The values of the electric field strength E_c at the conductor surface, available from the previous iteration, are compared with the value E_0 given by Peek's formula. The departure between the computed values of E_0 and E_c are used to retouch ρ_+ and ρ_- appropriately for convergence to speed up. The authors compared the results obtained by the described procedure with available experimental data. At first, a DC unipolar line model is considered and a fairly good agreement (errors lower than 20%) is obtained, for zero and nonzero wind speed, between calculated and measured lateral profiles of the electric field and current density on the ground. Also a pair of full-scale bipolar lines was simulated and the results were compared with available data. A rather similar agreement is again obtained, with the specification that the errors were in those cases higher than 20%, a reasonable expectation in view of the supplement of uncertainties arising when a real installation, exposed to uncontrollable ambient conditions, becomes the system under test.

In a further CSM-based procedure by Abdel-Salam, El-Mohandes, and El-Kishky (1989), both unipolar and bundle full-scale transmission lines are the examples taken into account. The authors can avoid use of DH and at last compare the given *I–V* characteristics with those of a previous paper by Abdel-Salam, Farghally, and Abdel-Sattar (1983) conversely affected by DH. The compared curves differ significantly, but those predicted by the CSM are closer to the experimental data.

(B3) Finite-Difference Method (FDM)   Bouziane et al. (1994) made use of a finite-difference algorithm for simulating the 2D charge problem related to noncoaxial cylinders (paraxial geometry). The method described is general and can be applied to 3D problems under DC conditions. This method considers an incremental section of a fluxtube, as depicted in Figure 2.8. The variations of the field quantities on each equipotential cross section are assumed to be so sufficiently small that all these quantities can be written as functions of the position *l*. The linear parameter *l* runs along the median fieldline, starting from the source. With reference to two equipotential cross

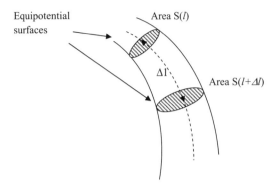

FIGURE 2.8   Sketched fluxtube section (full line) with dashed median fluxline. Origin for the function l of position on the emitting surface (out of sight).

sections mutually distanced by the quantity Δl, invoking current continuity gives $J(l) \, S(l) = J(l + \Delta l) \, S(l + \Delta l) = J(0)S(0)$, where $J(0)$ is assumed to be known. Once KH is adopted and a potential difference ΔV is prefixed, the length of the incremental section Δl for a given fluxtube can be derived because current density, mobility, and charge density are known at an initial point. The new equipotential surface is obtained by connecting these points. By introducing the termed dilation function $f(l) = S(l)/S(0)$, which takes into account wall dilation/restriction of a generic fluxtube as l changes, a suitable iterative solution for the current density $J(l)$ and electric field $E(l)$ can be implemented. The process runs until the grounded electrode is reached. The analysis is applied to both the ionization and drift regions of the electrode spacing. In particular, an additional electronic component is to be taken into account for the active zone other than the ionic one, the latter uniquely filling the drift region. As regards the paraxial geometry, a good agreement between simulated and measured J- and E-profiles is appreciable. In view of the debate on DH, it is of interest to remark that the given Poissonian paths show a smaller curvature than the Laplacian ones at lower voltages, a trend that reverses at higher voltages.

By a different method, Ohashi and Hidaka (1998) assigned to a wire-plane configuration a cardioid-like current distribution law of the type $J_e(\theta) = J_e(\theta = 0)(1 + \cos \theta)^n$ at the wire, with $n > 0$ being a voltage-dependent parameter. A comparison between simulated and measured data furnished by Hara et al. (1982) leads to the same conclusion discussed above. Moreover, the normalized current density profiles are almost independent of the applied voltage, with a performance-sustaining WL to be at least applicable to wire-to-plane coronas.

(B4) Hybrid Approaches   The FEM is also used by Davis and Horburg (1986) to solve Eq. (2.27) once a charge-density distribution ρ_0 is

assigned on the electrode surface. This method is combined with the Method of Characteristics (MOC) to calculate the charge-density distribution ρ in the gap. In the MOC, the unipolar charge-drift formula (2.10) is involved to give the space-charge density along some so-called characteristic lines, which identify with fieldlines and ion trajectories. There is a similar approach, one adopted by Adamiak (1994), in which the FEM is replaced with the Boundary Element Method (BEM) to solve Eq. (2.26). This expedient allows some interfacing difficulties, arising when the MOC is adopted, to be circumvented. Also additional requirements, such as using an approximate low-order polynomial for calculating the potential over a finite element and introducing special techniques to update the charge distribution on the characteristic lines, favorably drop. Because the BEM only needs discretization of the boundary, the sizes of the matrix are managed and, in turn, the computational time becomes significantly reduced. This coupled BEM-MOC method has been initially applied to the wire-duct configuration. The calculated V–I characteristic for a single wire is in good agreement with the corresponding experimental data. Moreover, the author observed how much different the simulated Laplacian and Poissonian patterns are and in essence concluded that DH is inherently invalid.

An improved version of the above procedure was later applied by Adamiak and Atten (2004) to the point-to-plane configuration with differently sharpened rods. The simulated results, expressed in terms of V–I characteristics, agree with the experimental curves when the ion mobility is set equal to $2.3 \times 10^{-4}\,\mathrm{m^2/(V \cdot s)}$. The theoretical and experimental current density distributions on the ground plane shows small differences, both closely fitted by a WL with an exponent equal to 4.82. The more the electrode sharpens, the less the estimation is accurate, probably due to the inability to correctly reproduce a very large field gradient near the electrode tip.

2.5 SPECIAL MODELS

2.5.1 Drift of Charged Spherical Clouds

In Jones et al. (1990), the unipolar charge drift is governed by Eq. (2.10) as a means of providing an analytical alternative to WL. The discharge is envisioned as being made by a succession of noninteracting, self-expanding spherical charge clouds while drifting in a uniform field of average value V/a. (see Figure 2.9). At first, the arbitrary assumption of a uniform field in a point-to-plane configuration legitimizes some objections to the model soundness.

The current distribution on the plane is assumed to be the result of charge accumulation when the spherical clouds impact the plane. Indeed, the spheres are thought of as passing through the plane without suffering distortion. The

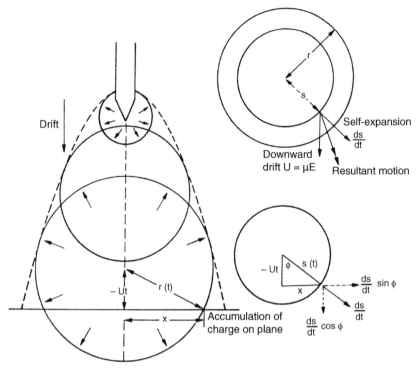

FIGURE 2.9   Point-plane configuration. Self-expanding sphere model by Jones et al. (1990). [Reproduced with permission from *Journal of Physics D: Applied Physics.*]

charge velocity at the plane is composed of $v = kV/a$ and ds/dt, the latter governing the cloud self-expansion. Therefore,

$$J(x) = \frac{\varepsilon v}{3k} \left[2\ln\left(1 + \frac{k\rho_0}{\varepsilon}t\right) + \left(1 + \frac{k\rho_0}{\varepsilon}t\right)^{-1} \right]_{t_1(x)}^{t_2(x)} \tag{2.33}$$

is the given formula for calculating the current distribution at the plane during the unperturbed passage of each sphere through the plane. Notation ρ_0 stands for the charge density of the spherical cloud at the time $t = 0$; $t_1(x)$ and $t_2(x)$ are the time instants for which the distance of the point of abscissa x from the center of the sphere is equal to its radius.

The rapid succession of spheres emanating from the point results in charge accumulation at each location x at the collector, thus giving rise to a steady current-density profile. Figure 2.10 shows a comparison between theoretical and experimental curves. The theoretical values of $J(x)$ ostensibly give on-axis overestimations, which may be derived from *a priori* unpredictable concurring causes. As it stands, this model is suitable to explain the second-order

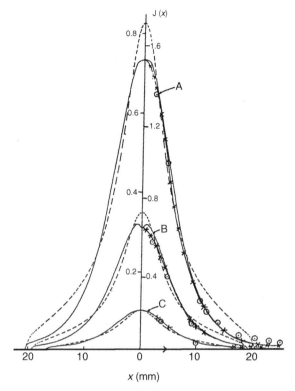

FIGURE 2.10   Point-plane configuration. Comparison between theoretical (see Figure 2.9) and experimental curves. —, Measured; ---, calculated (Jones et al. (1990)); × Warburg law; ⊙, Sigmond (1986). A: Current $I = 150\,\mu A$, $r_0 = 15.9$ mm, $\rho_0 = 0.00225\,Cm^{-3}$. B: Current $I = 65\,\mu A$, $r_0 = 15.0$ mm, $\rho_0 = 0.0171\,Cm^{-3}$. C: Current $I = 20\,\mu A$, $r_0 = 13.6$ mm, $\rho_0 = 0.0013\,Cm^{-3}$. [Reproduced with permission from *Journal of Physics D: Applied Physics.*]

detail represented by the remote cutoff of the experimental curve, whereas the central dimple holds unexplainably (see Section 5.4.1 for detailed experimental considerations on these curve features).

2.5.2 Graphical Approach

Also of some interest is a method based on a simple geometrical construction describing the asymptotic ion flow for a rod-plane configuration [see Amoruso and Lattarulo (1997)]. This graphical procedure leads to a closed form for the current density $g(x)$ at the plane which turns out to be an attractive alternative to WL. If the grounded plane is removed, a radial ion injection, somewhat evoking the previously claimed spherical expansion by Jones et al. (1990), is assumed because the actual injecting source of ions is virtually replaced with an isolated spherical spot. When the plane is restored, the straight trajectories

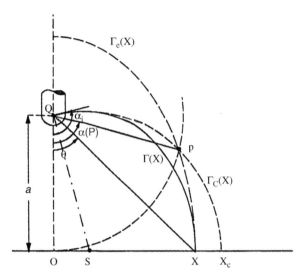

FIGURE 2.11 Point-plane configuration. Basic constructional details showing the α_i–x correlation. $\Gamma(x)$ is an envelope of adjacent elliptic arcs whose upper foci lie along the segment SQ. The curve segment $\Gamma_c(x)$ emerges from Q with inclination (injecting) angle α_i; $\Gamma_e(x)$ impacts orthogonally the ground at the abscissa x. According to a bipolar representation, S runs on the x-axis as α_i and, in turn, $\Gamma(x)$ change. [Reproduced from Amoruso and Lattarulo (1997) with permission from *Journal of Electrostatics*.]

bend owing to the plane influence. Both Deutsch's and Kaptzov's hypotheses concur in ultimately furnishing the function $g(x)$, which derives from preserving current continuity along any elemental fluxtube.

With reference to Figure 2.11, the graphical procedure consists of the following salient points:

- As soon as the ion flow is radially injected into the drift region all around a spherical source of unspecified small radius, the trajectories assume an initially circular geometry according to a sphere-to-plane Laplacian pattern.
- The inactive regions of the rod concur with the plane in giving rise to a different electrostatic influence during the time of flight; the ultimate result is that the destination point is reached when any given circular trajectory progressively becomes elliptic, thus rather pertaining to a Laplacian pattern in a rod-to-plane gap.
- As indicated elsewhere [see, in particular, Ciric and Kuffel E (1983)], the charge injection obeys the cardioid law:

$$g(\alpha_i) = g(\alpha_i = 0)\frac{1 + \cos\alpha_i}{2} \tag{2.34}$$

- Γ_c and Γ_e intersect on the point P lying on a circle, of radius equal to the elevation a in Figure 2.11, concentric with the source. Note that this circle is consistent with a spherical expansion originated at Q. The curve Γ_e is a quarter of an ellipse whose foci lie on Q and its mirror image. The segment SQ, which is perpendicular to the injecting direction identified by the angle α_i, is the locus of the foci of adjacent elliptic arcs whose envelope is Γ. The end foci S and Q correspond, respectively, to Γ_c emitted from Q with inclination angle $\alpha(P)$ and Γ_e impacting orthogonally the plane on the abscissa x.

- Basic geometrical considerations lead to establish that

$$
\alpha_i =
\begin{cases}
\tan^{-1}\left[\dfrac{2\sin\alpha(P)}{2\cos\alpha(P)-1}\right], & \alpha(P) \leq 60°, \\[3mm]
\pi + \tan^{-1}\left[\dfrac{2\sin\alpha(P)}{2\cos\alpha(P)-1}\right], & \alpha(P) > 60°
\end{cases}
\tag{2.35}
$$

with

$$
\alpha(P) = \cos^{-1}\left[\sqrt{1+\left(\frac{x}{a}\right)^2} - \left(\frac{x}{a}\right)^2\right]
$$

according to a spherical charge emission law.

- The resolving normalized formula for the planar current density profile becomes

$$
g(x) = g(x=0)\frac{1+\cos\alpha_i}{4}\frac{\sin\alpha_i}{\sin\alpha(P)}(2-\cos\theta)\frac{2-\cos\alpha(P)}{5-4\cos\alpha(P)} \tag{2.36}
$$

Note that $\alpha(P)$, α_i, and θ are functions of x, so $g(x)$ vanishes when $\theta = \theta_M = 60°$, namely, when $\alpha(P) = \alpha_i = 180°$ and $x = a\tan\theta_M = x_M$.

Figure 2.12 shows the resulting good agreement, better than one involving WL, between the theory under examination and experimental data. Finally, Figure 2.13 shows a discrete number of trajectories obtained by making use of the described graphical technique. Note that when β vanishes, then the envelope approaches a circle arc, throughout, whereas the curves with $\alpha_i > 90°$ tend to fit quarters of ellipses.

In an improved version of this procedure [see Amoruso and Lattarulo (2001)], the actual source of the ion flow remains unchanged, thus represented by a spherical spot irrespective of the real configuration of the rod terminal. Instead, the previous injecting cardioid polar law is replaced by a

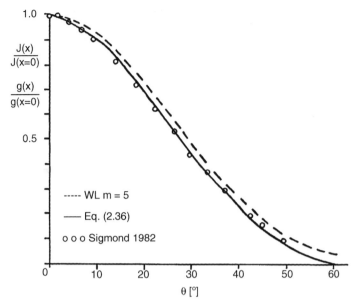

FIGURE 2.12 Point-plane configuration. Normalized theoretical profile of the current distribution compared with Warburg's cosine law and experimental data. [Reproduced from Amoruso and Lattarulo (1997) with permission from *Journal of Electrostatics*.]

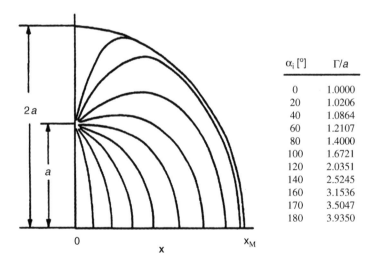

FIGURE 2.13 Point-plane configuration. Selected sequence of theoretical trajectories $\Gamma(\alpha_i)$ with $\alpha_i = \alpha_i(x)$ according Eq. (2.41). *Inset*: $\Gamma(\alpha_i)$ lengths for some values of α_i. The closures of the bidimensional representation are the rod central axis and the planar x-axis (boundaries at fixed voltages), in conjunction with $\Gamma(x=0)$ and $\Gamma(x_M)$. Specifically, $\Gamma(x=0)$ is the on-axis line; $\Gamma(x_M)$ is a quarter of an ellipse of main semiaxes $2a$ and x_M (foci centered in the active spot and its image). [Reproduced from Amoruso and Lattarulo (1997) with permission from *Journal of Electrostatics*.]

sophisticated law consisting of the superposition of a pair of canonical functions, notably the limacon of Pascal $(f_1 + f_2 \cos \alpha_i)$ and two leaf-like lobes [see details next in Eq. (2.37)], each composing a canonical n-leaved rose. The very final purpose consists in presumably imposing at the source generalized boundary conditions, thus more consistent with second-order features that deliberately alter the simple cardioid law. These are discernible with reference to distanced observations, namely, those appearing in the form of a central dimple and remote cutoff on the current distribution at the collector. Accordingly,

$$f(\alpha_i) = \begin{cases} f_1 + f_2\cos \alpha_i + \sum_{k=1,2} R_i \sin(n_k\alpha_i), & \alpha_i \leq \alpha_{iM} \\ 0, & \alpha_i > \alpha_{iM} \end{cases} \qquad (2.37)$$

where α_{iM} is the injection angle corresponding to θ_M and R_i is a constant. The normalized planar corona current distribution reads now

$$\frac{J(x)}{J(x=0)} = f(\alpha_i) \frac{\sin \alpha_i}{x} \frac{d\alpha_i}{dx} \frac{1}{[\alpha_i/x \cdot d\alpha_i/dx]_{x \to 0}} \qquad (2.38)$$

Figure 2.14 shows the normalized density distributions at the plane obtained by the improved procedure. A set of experimental data from Allibone

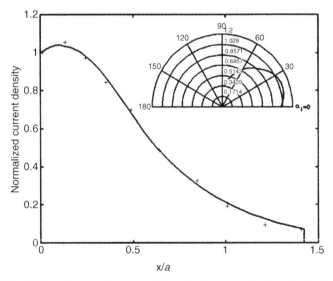

FIGURE 2.14   Point-plane configuration. Normalized current density distribution according to Eq. (2.38). x-axis: Normalized to the point elevation a. *Inset*: According to Eq. (2.37). Crosses: Experimental data from Allibone et al. (1993); $f_1 = 0.25$; $f_2 = 0.75$; $\alpha_{iM} = 84.35$; $R_1 = 0.06$; $n_1 = 2$ $R_2 = 0.09$; $n_2 = 4$. [Reproduced from Amoruso and Lattarulo (2001) with permission from *Journal of Electrostatics*.]

et al. (1993) is involved for the sake of comparison. The point charge source, simulating the active tip in the pin-plane case, has been assumed later in Ieta, Kucerovsky, and Greason (2005) to calculate the electric field and current density profiles on the planar collector. A Warburg-like cosine formula with an exponent equal to 6 is so obtained for the current density, thus of the type derived by Usynin (1966) and Walsh, Pietrowskj, and Sigmond (1984) for the wire-plane configuration. As the proposed method neglects space charge, the authors ascribe to this deficiency the raised departure from WL.

2.6 MORE ON DH AND CONCLUDING REMARKS

Although the aim of this review is not to be exhaustive, substantial arguments have nonetheless been given to illustrate, in particular, how the debate revolving around the DH is still open. There is a surprising, subtle reason that makes the above remark of prominent interest in the context of this book. Some advice on this concern will be given at the end of this closing section. Suffice it to say for now that estimating the DH validity is an exercise still under way and that the involved investigations persist to contradict each other. For example, a numerical procedure suitable for a bipolar configuration has been recently implemented by Li et al. (2010). The numerical results are compared with experimental data furnished by using either a small-scale laboratory model or a real HVDC bipolar transmission line installation. The authors confirm that DH introduces considerable errors that tend to increase with the severity of the corona discharge in itself (nearly according to a linear law) and with the applied voltage (in view of field distortion). These errors are slightly lower if bundle conductors are adopted, and they are reduced when the bundle number increases. On the other hand, Yang, Lu, and Lei (2008) used the DH for analyzing the behavior of four schemes of double-circuit transmission lines. The analysis takes into account the different conditions of each pole with respect to its own onset voltage (also dependent on polarity). Validation of the theoretical results is made by comparison with experimental data, and an observed performance is that when the difference between the corona-onset electric fields of the positive and negative poles is considered, then the theoretical results are closer to the corresponding experimental data.

In the authors' opinion, the tedious debate on a matter as important as DH is being turned around in precisely the wrong way. The efforts made up to now to judge HD are estimated to be unfruitful merely because they are based on questionable premises attributable to the adopted computational tools. In fact, the available investigation may have tried its best, but it seems it did the wrong thing by using decoupled models. On the contrary, a successful model is claimed in Chapter 4, later corroborated in Chapter 5, whose coupled character shows to be totally at one with DH. In the light of this incidental issue, it might be worth interpreting DH as a physical law rather than as a hypothesis still subject to validation. Therefore, it is DH—let us call it that again—which should be regarded as a reference for model validations, not the contrary.

APPENDIX 2.A: WARBURG'S LAW (WL)

In 1899, Warburg first published the empirical law, later called by his name, governing the plane distribution of the dc corona current for point-plane gaps with $a \leq 3$ cm in ambient air [see, for example, Jones (1992)]. It postulates that the current distribution $J(x)$ beneath the point at a radial distance x from the axis is independent of the endpoint geometry and can be formulated as follows:

$$J(x) = J(x = 0)\cos^m\theta, \qquad \theta = 60° \tag{2.A.1}$$

Here, $\theta = \tan^{-1}(x/a)$ (see Figures 2.2–2.4); $J(x=0)$ is the central peak value of the current distribution (the latter intended to be expressed as a quadratic function of the applied voltage); $m = 4.82$ for positive corona and 4.65 for negative corona. For negative discharges, the law is valid in the voltage range from the corona inception to a little below breakdown; for positive discharges, the law is valid as long as vivid streamers are absent. WL also holds for an extended range of air humidity, irrespectively of the voltage polarity [see Goldman et al. (1988)]. Moreover, the experiments described by Allibone and Saunderson (1989) show that the initial limitation to gaps ≤ 3 cm for the WL validity can be extended to a few meters.

Two important features of this law, which seem to tacitly agree with each other, can be highlighted briefly:

- The current distribution $J(x)$ is substantially irrespective of the tip geometry, in the sense that specific phenomena occurring in/surrounding the active region attached to the tip tend to mask its minute geometric features.
- Warburg's law safely holds only when air is the filling gas. The remarkable departures in the detected profiles when air, nitrogen, and oxygen were individually used as gaseous media were ascribed (Goldman et al., 1988) to the fact that discharges in nitrogen and oxygen are completely different from those in air.

The most important criticism of this law regards the lack of a physical interpretation of the discharge, specifically whether the cosine law be essentially dependent on the drift region or on the ionization region. Incidentally, it is worth bearing in mind the advantage of framing this question in the leading subject of Chapters 4 and 5.

As regards the curve anomalies represented by the central dimple and the remote cutoff, clearly shown in Figure 2.14, several physical explanations have been proposed. For example, Goldman, Selim, and Waters (1978) state that the small-amplitude, rapid on-axis fluctuations of the discharge surrounding the tip could be a reason for the dimple. Davies et al. (1987) discuss various possible causes for the cutoff and conclude that the presence of two different ionic species could explain the phenomenon for a certain range of current.

APPENDIX 2.B: BIPOLAR IONIZED FIELD

In the case of a unipolar corona, the polarity of ionic species is coincident with that of the active electrode. For bipolar configurations (see Figure 2.7), ions of both polarities are injected into the common gap; positive (negative) ions dominate in the region between the positive (negative) electrode and ground. The field in the positive and negative regions may still be described by Eqs. (2.1)–(2.3). In the region between the two electrodes, ionic species of both polarities are present. The field in the bipolar region is believed to obey the following set of formulas:

$$\nabla \cdot \mathbf{E} = \frac{\rho_+ - \rho_-}{\varepsilon_0}, \tag{2.B.1}$$

$$\mathbf{J}_+ = \rho_+ \left[k_+ \mathbf{E} - \left(\frac{D_+}{\rho_+} \right) \nabla \rho_+ + \mathbf{w} \right], \tag{2.B.2}$$

$$\mathbf{J}_- = \rho_- \left[k_- \mathbf{E} + \left(\frac{D_-}{\rho_-} \right) \nabla \rho_- - \mathbf{w} \right], \tag{2.B.3}$$

$$\nabla \cdot \mathbf{J}_+ = -\frac{R_i \, \rho_+ \rho_-}{e}, \tag{2.B.4}$$

$$\nabla \cdot \mathbf{J}_- = \frac{R_i \, \rho_+ \rho_-}{e} \tag{2.B.5}$$

Rearranging Eqs. (2.B.2)–(2.B.5) ultimately gives

$$\nabla \cdot \left[(k_+ \mathbf{E} + \mathbf{w}) \rho_+ - D_+ \nabla \rho_+ \right] + \left(\frac{R_i \, \rho_-}{e} \right) \rho_+ = 0 \tag{2.B.6}$$

$$\nabla \cdot \left[(-k_- \mathbf{E} + \mathbf{w}) \rho_- - D_- \nabla \rho_- \right] + \left(\frac{R_i \, \rho_+}{e} \right) \rho_- = 0 \tag{2.B.7}$$

Here, R_i is the ionic recombination coefficient, w is the wind speed, and D_+ and D_- are the thermal diffusion coefficients for positive and negative ionic species, respectively. The remaining quantities involved have already been introduced somewhere else. When ρ_+ (or ρ_-) is set equal to zero, the set of reduced equation governs the negative (positive) unicharge case. In order to alleviate the computational effort, the following simplifying approximations, especially suitable to the usual case study of an overhead power line, can be introduced:

- The ionic mobilities k_+ and k_- are constant and, in general, mutually different quantities.
- The thickness of the ionized layer around the injectors is neglected in comparison to the electrode spacing.
- For applied voltages above the corona onset level V_T, the surface potential gradient on the active electrode remains equal to a threshold E_0, related to V_T, even if $V > V_T$ is theoretically changed at will (Kaptzov's hypothesis).
- Corona activities and wind velocity are steady phenomena; the presence of towers, conductor sagging, and stranding is neglected.

CHAPTER 3

INTRODUCTORY SURVEY ON FLUID DYNAMICS

3.1 INTRODUCTION

For the benefit of those that are not familiar with fluid dynamics, some basic subjects relevant to the subsequent chapters can be found in the following sections. Mention of the ion flow is gently made on cue so as to somehow smooth the way for the reader. In fact, this chapter is especially introductory to a joint theory, involving the electromagnetic counterpart, which will be treated extensively in Chapter 4. Forces of electrostatic (or magnetic) nature impressed to the drifting ions are represented here in the typically general form of body forces. It should be remarked in advance that a conceptual difficulty can at first arise in assimilating the drift of charge-bearing molecules to the flow customarily treated in fluid dynamics. In the former case, only a fractional amount of the totality of molecules takes part in the motion which, hence, assumes the character of a continuously obstructed drift because of the massive presence of buffer neutrals. In the latter case, instead, all the molecules composing the fluid are subject to the body force and concur to the fluid motion. The raised differences between the motions under examination have, as a matter of fact, persuaded previous investigators to restrict fluid-dynamic-related assumptions only to convection currents. In other words, the drift model so far appears substantially deprived of some crucial fluid dynamic conditions. A paradigmatic example regards ion drifts

Filamentary Ion Flow: Theory and Experiments, First Edition. Edited by Francesco Lattarulo and Vitantonio Amoruso.

disturbed, say, by atmospheric wind. The adopted treatments safely take into account some essential fluid dynamic performances only for the wind-originated convection component of the current while the drift component is left unaffected. As unfortunately and invariably expected, the given theoretical predictions are subject to unrelenting debate. Just to give some sense of the theoretical difficulties that have arise in the past, and, hence, of the ensuing historical controversies, in introducing basic notions of nonrelativistic continuum mechanics and thermodynamics for electrodynamic models to be ultimately formulated, it is rewarding to read Chapter 8 of an early book written by Penfield and Haus (1967) and the paper authored by Felici (1978). A careful attempt in line with a coupled modeling was later made by Dulikravich and Lynn (1997a,b). As noticed beforehand, the subject matter will be reconsidered in great detail in Chapter 4, where the notions of reduced mass and charge are introduced in tandem with an unprecedented coupling model for drifting ions. Unless otherwise specified, this chapter tacitly takes account, throughout, of the book by Warsi (1999), to which the interested reader is invited to refer for more extended treatment.

3.2   CONTINUUM MOTION OF A FLUID

The moving fluid of considerable interest in the present treatment is air with ambient parameters largely distant from those pertaining to the critical state. Therefore, using the perfect-gas model turns out to be a largely permissible practice without loss of accuracy. Interpreting such a gaseous medium according to a statistical or continuum concept is of fundamental importance in the analysis of the fluid motion. Therefore, a pair of representative micro- and macroscale lengths needs to be compared; respectively, they are the mean free path λ of the gas and the characteristic length a_p of a fluid particle. The random character of the molecule motion and the related mutual collisions are microscopic performances importantly represented by λ. Otherwise, the so-called fluid particle is an elementary quantity of matter that macroscopically moves with the fluid. Discerning appropriately between the two above-mentioned approximate idealizations is made permissible by introducing Knudsen's number

$$K_n = \frac{\lambda}{a_p} \qquad (3.1)$$

The rationale consists, according to Eq. (3.1), of evaluating the above dimensionless ratio and, then, in verifying whether the two-sided bound condition $0 \leq K_n \leq 1$ is satisfied or not. In the affirmative case, the medium behaves as a continuum; otherwise, it needs to be formulated according to statistical mechanics. Incidentally, consider that a_p is rather representative of a material line element associated with one of the longitudinally continuous stream filaments composing a transversally discontinued multichanneled flow pattern. As will be extensively treated in Chapter 4, such a pattern is appropriate to a subsonic ion flow, and the cross-sectional dimension of each filamentary channel is argued to shrink in order to become as narrow as it can physically be, namely, to approach the molecular diameter (order of some unities of 10^{-10} m) of the gas. Of course, nothing prevents us from extending Eq. (3.1) to the more general example of a compact flow, thus composed of contiguous elementary stream tubes (also referred to as fluxtubes) of an unspecified small cross section whose walls dilate/contract along their extension. In any case, a usual ion-flow crossed interelectrode spacing is such that the actual length L of a generically curved fluxtube attains the order of 10^{-1} to 10 m. Therefore, setting $\lambda = 6 \times 10^{-8}$ m for the gaseous filler, appropriately replacing the generic notation a_p with ΔL in Eq. (3.1), and using the above-discussed rationale gives

$$\infty > L \gg \Delta L \geq 10^{-5}$$

The substantial result is that it is permissible to treat a fluxtube's elementary segment of length $\Delta L \geq 10^{-5}$ m as the physical counterpart of the mathematically infinitesimal length dL involved in the analysis of a continuum fluid. It is easy to realize how the above condition for ΔL is not severe at all for our practical applications. This means that no difficulty will arise in assigning the dimensions of a fluid particle that is demanded to appear large-sized in comparison to each inner molecule but small-sized in comparison to the overall extension of the fluid. Note that the above observations are quite irrespective of fluid dynamics since Eq. (3.1) involves velocity-independent quantities. Incidentally, tacit admission is made that λ is related to a gas in thermal equilibrium, in which state whether the fluid is at rest or subject to motion turns out to be an unimportant detail. However, a subsidiary attention could be paid to the case of an unphysical gas with zero viscosity μ or, similarly, to that of a physical gas subject to specialized dynamics as those illustrated later in Sections 3.10 and 3.11. Under these circumstances, the gaseous filler behaves as a perfect continuum with $K_n = 0$. This result derives from setting $\mu = 0$ and considering that $\lambda \propto \mu$ represents a linear relationship governing the molecular model.

3.3 FLUID PARTICLE

From here on, the fluid will be considered as a continuum, which implies that the flow can be formally expressed as a continuous time-dependent transformation of the same 3D Euclidean space. Accordingly, an elementary material element of the continuum, previously indicated as a fluid particle, generally happens to be subject to longitudinal and angular deformations as it moves with the flow. In other words, the volume dV and the closed bounding surface dS of a particle of constant mass $\rho_m dV$ are both obliged to change. The notation ρ_m has been adopted for the mass density $\rho_m = nm_0$, which, therefore, is related to the varying number density n and the mass $m_0 = 4.8 \times 10^{-26}$ kg of a single molecule. It should be noted that the mass conservation assumed for the moving particle refers to the same molecules (in number and identity) collectively in motion. Therefore, the fluid particle identifies with a given closed subsystem whose bulk motion differs from the case in which only a fractional amount of the fluid's molecules can move in the presence of a buffer mass (see the introductory section, Section 3.1). The latter is composed of obstructing motionless molecules with individual masses equal to and/or different from m_0. In this situation, the notion of a moving particle needs to be carefully reexamined since the particle is now composed of the same number of molecules whose motionless fractional amount is continuously replaced during the motion. This aspect of the problem will be dealt with and resolved in Chapter 4, with specific reference to ions drifting in a neutral gas when an exogenous electric field is applied.

3.4 FIELD QUANTITIES

The geometrical and material quantities dS, dV, ρ_m, and n introduced in the above section are single-valued spatiotemporal functions. Accordingly, they are typically classified as fields, even in consideration of the relative extension of the common space domain occupied by the moving fluid. Indeed, a number of additional physical and kinematic functions with the mathematical attributes of scalar, vector, or tensor fields are implied in the analysis of a flow. We can identify such fields with the generic tensor $\mathbf{\Gamma}$ since, in particular, a tensor with order zero or one is a scalar or a vector, respectively. According to an Eulerian framework, $\mathbf{\Gamma} = \mathbf{\Gamma}(\mathbf{r}, t)$ is fully identified by the particle position $\mathbf{r}$ and the time t (fixed reference frame), so that the local rate of change of $\mathbf{\Gamma}$, when the fluid particle passes through the point $\mathbf{r}$ at the instant t of time, can be evaluated consequently in a straightforward and appropriate way. However, preference is often expressed for the alternative Lagrangian formulation in the co-moving frame, thus based on following a given particle (material element of fixed identity) during the motion [see also Bennet (2006)]. This approach calls into play the substantial (or material) derivative, whose symbolic notation

$$\frac{D\mathbf{\Gamma}}{Dt} = \left(\frac{\partial}{\partial t} + \mathbf{v} \cdot \text{grad} \right) \mathbf{\Gamma} \tag{3.2}$$

differently written as

$$\frac{D\Gamma}{Dt} = \frac{\partial\Gamma}{\partial t} + (\text{grad }\Gamma) \cdot \mathbf{v} \tag{3.3}$$

involves the absolute velocity $\mathbf{v} = d\mathbf{r}/dt$.

Applying the notion of substantial derivative to a material parcel (intended as being composed of a large amount of particles) of volume $V(t)$ bounded by the closed surface $A(t)$ whose outwardly directed unit vector is $\mathbf{n}$ will ultimately yield

$$\frac{D}{Dt} \int_{V(t)} \Gamma \, dV = \int_{V(t)} \left[\frac{\partial\Gamma}{\partial t} + \text{div}(\Gamma \cdot \mathbf{v}) \right] dV \tag{3.4}$$

or

$$\frac{D}{Dt} \int_{V(t)} \Gamma \, dV = \int_{V(t)} \left[\frac{D\Gamma}{Dt} + \Gamma \text{div } \mathbf{v} \right] dV \tag{3.5}$$

Equations (3.4) and (3.5) are two alternative formulations allowed by the so-called transport theorem for the derivation under the sign of integration. Indeed, this theorem is a direct consequence of the divergence (or Gauss') theorem $\int_{A(t)} \Gamma \cdot \mathbf{n} \, dA = \int_{V(t)} \text{div}(\Gamma \cdot \mathbf{n}) \, dV$.

3.5 CONSERVATION LAWS IN DIFFERENTIAL FORM

3.5.1 Generalization

Irrespective of the physical nature and mathematical formulation of the Γ-field, the equation

$$\frac{D}{Dt} \int_{V(t)} \Gamma \, dV = -\int_{A(t)} \Phi \, dA + \int_{V(t)} \Psi \, dV \tag{3.6}$$

formalizes the equality between the rate of change of Γ within a material parcel as it moves with the flow [left-hand side of Eq. (3.4)] and the rate of Φ and Ψ transfers (right-hand side of the same equation). Even the unspecified agents Φ and Ψ are to be assumed as tensors, in particular Φ being one order higher than Γ and Ψ. This is because the latter pair of tensors is involved in volume integrals, whereas Φ is the integrand of a surface integral. Again using the divergence theorem for the first term on the right-hand side of Eq. (3.6), rearranging it with Eq. (3.4), and invoking the validity of the given equality for arbitrary volumes will finally give

$$\frac{\partial \mathbf{\Gamma}}{\partial t} + \text{div}(\mathbf{\Gamma v}) = -\text{div}\,\mathbf{\Phi} + \mathbf{\Psi} \qquad (3.7)$$

Equation (3.7) represents a possible generalization of the conservation law in the differential form. A careful choice in the involved tensors $\mathbf{\Gamma}$, $\mathbf{\Phi}$ and $\mathbf{\Psi}$, in which case the transfer tensors $\mathbf{\Phi}$ and/or $\mathbf{\Psi}$ could also be demanded to vanish, allows the law under examination to be specified for any given physical application.

3.5.2 Mass Conservation

Characteristic of mass conservation is setting $\mathbf{\Gamma} = \rho_m$ and $\mathbf{\Phi} = \mathbf{\Psi} = 0$. Accordingly, Eq. (3.7) becomes

$$\frac{\partial \rho_m}{\partial t} + \text{div}(\rho_m \mathbf{v}) = 0 \qquad (3.8)$$

otherwise written

$$\frac{\partial \rho_m}{\partial t} + \mathbf{v}\,\text{grad}\,\rho_m + \rho_m\,\text{div}\,\mathbf{v} = 0 \qquad (3.9)$$

by virtue of a tabulated identity for the divergence of a product. Incidentally, Eq. (3.8) is also referred to as the continuity law because ρ_m and $\mathbf{v}$ are tacitly assumed to be continuously differentiable field quantities. Therefore,

$$\frac{D\rho_m}{Dt} + \rho_m\,\text{div}\,\mathbf{v} = 0 \qquad (3.10)$$

after remembering the general equation, Eq. (3.2). Equations (3.8) and (3.10) represent the same statement respectively expressed by Eulerian (fixed reference frame) and Lagrangian (co-moving reference frame) formulations. Note, in particular, that Eq. (3.10) descends directly from Eq. (3.5) since

$$\frac{D}{Dt}\int_{V(t)} \rho_m \, dV = 0 \qquad (3.11)$$

which tacitly advises how an observer linked up with a parcel of volume $V(t)$ and mass $\int_{V(t)} \rho_m \, dV$ perceives the latter quantity as a conserved property.

3.5.3 Momentum Conservation

Equation (3.7) needs to be specified by using the following substitutions:
$\boldsymbol{\Gamma} = \rho_m \mathbf{v}$ (momentum per unit volume); $-\boldsymbol{\Phi} = \mathbf{T}$ (Cauchy stress tensor) and $\boldsymbol{\Psi} = \rho_m \mathbf{f}_m$ (body force per unit volume) where $\mathbf{f}_m$ is the body force per unit mass intended as an external force directly acting on the inner points of each fluid particle. The $\mathbf{f}_m$-field is a wide range one, thus involving the entire extension of the fluid or large amounts of it. This force can be originated by distanced or nearby sources of unspecified superficial or volumetric extension which, in the case of an ion flow, produce a background electric field. On the other hand, $\mathbf{T}$ is a stress tensor related, through the compact version of the Cauchy theorem $\mathbf{n} \cdot \mathbf{T} = \boldsymbol{\tau}$, to the external force $\boldsymbol{\tau}$ per unit surface enclosing the parcel. This stress originates from short-range molecular activities between contiguous particles; likewise, $\mathbf{f}_m$ acts internally to each parcel. Note that $\boldsymbol{\tau} = \boldsymbol{\tau}(\mathbf{n})$, in the sense that it depends on the shape assumed in a given instant by the outer surface of the particle, namely, on the orientation of the local unit vector $\mathbf{n}$. Both $\mathbf{f}_m$ and $\boldsymbol{\tau}$ are externally impressed to the particle under examination, while internal forces due to intermolecular interactions are set to zero. This assumption is directly consistent with the features of an ideal gas and indirectly consistent with the continuum model also adopted, in which case the influences among particles cancel out in a macroscopic reading of the system.

Therefore,

$$\frac{\partial(\rho_m \mathbf{v})}{\partial t} + \mathrm{div}(\rho_m \mathbf{v}\mathbf{v}) = \mathrm{div}\,\mathbf{T} + \rho_m \mathbf{f}_m \qquad (3.12)$$

which alternatively reads

$$\rho_m \frac{D\mathbf{v}}{Dt} = \mathrm{div}\,\mathbf{T} + \rho_m \mathbf{f}_m \qquad (3.13)$$

Note that Eq. (3.12) unfortunately embodies the dyadic $\mathbf{v}\mathbf{v}$ which contributes in making the physical interpretation of that formula rather unintelligible. This conceptual difficulty can be partially circumvented by considering that the acceleration $D\mathbf{v}/Dt$ can be expressed in a purely vector form. Accordingly,

$$\frac{D\mathbf{v}}{Dt} = \frac{\partial \mathbf{v}}{\partial t} + (\mathrm{grad}\,\mathbf{v}) \cdot \mathbf{v} = \frac{\partial \mathbf{v}}{\partial t} + \mathrm{grad}\left(\frac{v^2}{2}\right) - \mathbf{v} \times \mathrm{curl}\,\mathbf{v} \qquad (3.14)$$

by using a tabulated identity where the permissible rotational nature of $\mathbf{v}$ is manifested through the so-called vorticity term curl $\mathbf{v}$. Rearranging Eqs. (3.13) and (3.14) yields

$$\frac{\partial \mathbf{v}}{\partial t} + \text{grad}\left(\frac{v^2}{2}\right) - \mathbf{v} \times \text{curl } \mathbf{v} = \frac{1}{\rho_m} \text{div } \mathbf{T} + \mathbf{f}_m \qquad (3.15)$$

3.5.4 Total Kinetic Energy Conservation

In this case, setting $\Gamma = \rho_m e_t$; $-\mathbf{\Phi} = \mathbf{T} \cdot \mathbf{v} - \mathbf{q}_T$ and $\mathbf{\Psi} = \rho_m\,(\mathbf{f}_m \cdot \mathbf{v})$ yields

$$\frac{\partial(\rho_m e_t)}{\partial t} + \text{div}(\rho_m e_t \mathbf{v}) = \text{div}(\mathbf{T} \cdot \mathbf{v} - \mathbf{q}_T) + \rho_m(\mathbf{f}_m \cdot \mathbf{v}) \qquad (3.16)$$

The new quantities visible in Eq. (3.16) are the total kinetic energy per unit mass e_t expressed by the sum

$$e_t = e_i + e_k \qquad (3.17)$$

and the heat flux density vector $\mathbf{q}_T$. In particular, the specific internal energy e_i embodied in Eq. (3.17) and $\mathbf{q}_T = -\sigma_T \text{ grad } T$ take into account an amount of microscale energy modes, respectively connected to the molecular motion and outward heat transfer (σ_T and T denote the fluid's heat conductivity and absolute temperature). Instead, $e_k = v^2/2$ represents the kinetic energy per unit mass of the bulk motion, in which the simpler notation v^2 can replace the dot product $\mathbf{v} \cdot \mathbf{v} = |\mathbf{v}|^2$. Of course, all the above-mentioned specific quantities are on a unit-mass basis; they acquire a unit-volume basis after being multiplied by ρ_m. Substantially, Eq. (3.16) expresses a balance that involves the rate of increase of total energy per unit volume summed to inward convection of such energy (first and second terms on the left-hand side of that equation, respectively) and the sum of net heat flow, surface-force work, and body-force work (first, second, and third terms on the left-hand side, respectively).

The optional character of Eqs. (3.12) and (3.13) for the momentum conservation similarly comes up for the energy, provided that the identity

$$\frac{\partial(\rho_m e_t)}{\partial t} + \text{div}(\rho_m e_t \mathbf{v}) = \rho_m \frac{De_t}{Dt} \qquad (3.18)$$

is taken into account. In such a way, an alternative form for Eq. (3.16) simply consists of replacing the left-hand side of Eq. (3.16) with the right-hand side of Eq. (3.18).

3.6 STOKESIAN AND NEWTONIAN FLUIDS

The governing equations, (3.15) and (3.16), subject to the problem specification, allow the fluid motion under examination to be unambiguously resolved. For this class of applications, the above additional requirement implies that some simplifying assumptions are to be adopted for the fluid model and results in supplying some practicable formulation for the stress tensor $\mathbf{T}$. In a qualitative sense, the current properties of $\mathbf{T}$ depend on the deformation and rate of deformation of the particle moving with the fluid other than on the history/ future of the motion. In the present applications, the fluid is safely assumed to be independent of deformation and deprived of memory/precognition. By elimi- nation, the fluid properties will be sensitive only to spatiotemporal changes of the particle shape. Formally, a law of the kind $\mathbf{T} = \mathbf{T}(\mathcal{D})$ exists with $\mathcal{D} = \text{grad } \mathbf{v}$ designating the rate-of-strain tensor. If the equality $\mathbf{T} = p\,\mathbf{I}$ also holds at rest (in which case the only stress is p), then the fluid is classified as a Stokesian one. As regards the previous additional condition, p and $\mathbf{I}$ respectively denote thermo- dynamic pressure and unit tensor, the former equal to $-\partial e_i/\partial(1/\rho_m)$, the latter represented by a diagonal matrix with unitary entries.

Accordingly,

$$\mathbf{T} = (-p + \mu' \text{div } \mathbf{v})\mathbf{I} + 2\mu\mathcal{D} \qquad (3.19)$$

where μ and μ' are the viscosity and the second coefficient of viscosity, respectively, both in general dependent on the pressure p and temperature. Incidentally, a further admissible approximation consists of neglecting the dependence on p, which importantly implies that the fluid is identified as a Newtonian one. Under such circumstances, $\mathbf{T} = \mathbf{T}(\mathcal{D})$ becomes a linear law. The physical meaning of $\mathbf{T} = \mathbf{T}(\mathcal{D})$ and its reduction to a linear law—the latter feature is distinctive of Newtonian fluids—become immediately dis- cernible if use is made of a typically introductory shearing flow example. Let the flow be interposed between two indefinitely extended parallel plates when one of them moves with a given velocity with respect to the opposite one assumed, for simplicity's sake, at rest. Accordingly, the particles move relative to one another and the equality $\mathbf{T} = \mathbf{T}(\mathcal{D})$ can be reduced to the expanded form proposed in Marchi and Rubatta (1981):

$$\tau = \tau_1 + \mu\left(\frac{dv}{dx}\right) + \mu_1\left(\frac{dv}{dx}\right)^2 + \mu_2\left(\frac{dv}{dx}\right)^3 + \cdots \qquad (3.20)$$

since $\mathbf{T} = \tau$ and $\mathcal{D} = dv/dx$. Here, τ is the tangential shear stress, x is a coordinate that runs transversally to the flow and originates on the plate at rest, $v = v(x)$ is the velocity of each particle, and τ_1, μ_i $(i = 1, 2, \ldots)$ are unspecified quantities. A Newtonian fluid obeys a simplified version of Eq. (3.20) where all the terms vanish but the second one. This implies that there exists a linear relationship between τ and the velocity gradient dv/dx whose proportionality coefficient is exactly the viscosity μ. Air largely meets the conditions imposed by a Newtonian fluid, therefore, these will tacitly be assumed from now on.

Performing the divergence of Eq. (3.19) and introducing the ancillary tensor $\mathbf{S} = (\mu'(\text{div } \mathbf{v})\mathbf{I} + 2\mu\mathcal{D})$ gives

$$\text{div } \mathbf{T} = -\text{grad } p + \text{div } \mathbf{S} \tag{3.21}$$

Equation (3.21) is specifically required in Eq. (3.15) as it stands as well as in further manipulations of Eq. (3.16) aimed at describing momentum and energy conservations of viscous compressible fluids. Bear in mind that the conservation expressed by Eq. (3.8) for the mass is inherently independent of $\mathbf{T}$.

3.7 THE NAVIER–STOKES EQUATION

The momentum conservation can be expressed by rearranging Eqs. (3.13) and (3.21) to give the often cited Navier–Stokes equation

$$\rho_m \frac{D\mathbf{v}}{Dt} = -\text{grad } p + \text{div } \mathbf{S} + \rho_m \mathbf{f}_m \tag{3.22}$$

where

$$\text{div } \mathbf{S} = \text{div}(\mu'(\text{div } \mathbf{v})\mathbf{I} + 2\mu\mathcal{D}) = \text{grad } (\mu'(\text{div } \mathbf{v})) + \text{div}(2\mu\mathcal{D}) \tag{3.23}$$

Note that the composite terminology adopted above for Eq. (3.22) could be accounted for by introducing some useful simplifying conditions. In particular, if the viscosity coefficients μ and μ' in Eq. (3.23) hold uniform (homogeneity condition for the fluid mass), then Eq. (3.23) becomes

$$\text{div } \mathbf{S} = (\mu + \mu')\text{grad div } \mathbf{v} + \mu\nabla^2\mathbf{v} \tag{3.24}$$

after using some manipulations involving the identity $\nabla^2\mathbf{v} = \text{grad}(\text{div } \mathbf{v}) - \text{curl}(\text{curl } \mathbf{v}) = \text{div}(\text{grad } \mathbf{v})$. Therefore, Eq. (3.22) can be rewritten in the final form

$$\rho_m \frac{D\mathbf{v}}{Dt} = -\text{grad } p + (\mu + \mu')\,\text{grad div } \mathbf{v} + \mu\nabla^2\mathbf{v} + \rho_m\mathbf{f}_m \tag{3.25}$$

which represents the celebrated Stokes equation. Parenthetically, Eq. (3.25) results in the so-called Euler's equation if the viscosity is neglected, as is the case in practical situations except inside boundary layers. These filmy regions are attached to the walls and envelop inner bodies where the viscosity-dependent no-slip condition happens to apply. Assuming a zero rate of expansion for the moving particle, which is formally expressed by div $\mathbf{v} = 0$, causes Eq. (3.25) to become the equally celebrated Navier's equation

$$\rho_m \frac{D\mathbf{v}}{Dt} = -\operatorname{grad} p - \mu \operatorname{curl}(\operatorname{curl} \mathbf{v}) + \rho_m \mathbf{f}_m \qquad (3.26)$$

Incidentally, it is worth considering that under the above circumstances applied to a fluid particle, $D\rho_m/Dt = 0$ easily derives from Eq. (3.10). This implies that the constancy of both coefficients μ and μ' is inherent to Navier's equation (see also Section 3.9). Note further how $\nabla^2 \mathbf{v} = \operatorname{grad}(\operatorname{div} \mathbf{v}) - \operatorname{curl}(\operatorname{curl} \mathbf{v})$ inserted in Eq. (3.25) is now reduced to the second negative term alone.

3.8 DETERMINISTIC FORMULATION FOR e_t

After managing some straightforward mathematics involving Eqs. (3.16) and (3.21), we obtain

$$\frac{\partial(\rho_m e_t)}{\partial t} = \operatorname{div}(\mathbf{S} \cdot \mathbf{v} + \sigma_T \operatorname{grad} T - (\rho_m e_t + p)\mathbf{v}) + \rho_m (\mathbf{f}_m \cdot \mathbf{v}) \qquad (3.27)$$

Since e_t can be split according to Eq. (3.17), a careful reading of Eq. (3.27) leads us to realize how the work of the body force $\rho_m(\mathbf{f}_m \cdot \mathbf{v})$ specifically promotes the fluid's acceleration and e_k increases, whereas the heat flux only causes e_i to increase.

All the above equations governing mass, momentum and energy conservations, and related laws simplify further, provided that some important features are assumed for the fluid or flow. In hindsight, in the present survey, special attention is paid to the conditions of incompressibility and non-circularity simultaneously imposed onto a flow.

3.9 INCOMPRESSIBLE (ISOCHORIC) FLOW

3.9.1 Mass Conservation

A flow is by definition incompressible when $(1/\rho_m)(D\rho_m/Dt) = 0$, which means that the mass density ρ_m of each particle that moves through the flow field holds constant. Introducing this condition in Eq. (3.10) gives div $\mathbf{v} = 0$ and introducing this identity in Eq. (3.9) gives $\partial \rho_m/\partial t = 0$. Accordingly, a compressible fluid can behave as being incompressible during the motion, provided that the **v**-field is solenoidal or ρ_m is time-independent at any fixed point. These ambivalent definitions are alternative since Eq. (3.10) is only a different formulation of Eq. (3.9). The fact that ρ_m holds invariant in space and time implies that **v** and ρ_m are mutually independent, even under transient excitation conditions (at the time instants of field establishment/suppression); that is, strictly termed "unsteady conditions" are, as a matter of fact, interdicted. Under the described circumstances, the fluid mass is homogeneous in the complete sense that ρ_m, μ, and μ' are constant. Additionally, it is worth noting that the quantity $(1/\rho_m)(D\rho_m/Dt)$ can be expressed directly under the general form of the thermodynamic equation of state (see Appendix 3.A). The value of this equation will be appreciated later in this book as regards a special characterization (proposed here) of a material fluid that flows subsonically.

3.9.2 Subsonic Flow

Indeed, the only condition for the circumstances described in the above subsection to be physically permissible consists of imposing that the compressible fluid can flow without being subject to pressure changes. This occurs substantially if each fluid particle moves subsonically in an ambient delimited by adiabatic walls. The flow is subsonic if $v/c_s \leq 0.3$, where v is the modulus of **v** and c_s is the local speed of sound (c_s approaches the value of 343 m/s for air at 20 °C). The dimensionless ratio v/c_s, usually represented by the notation M, is the Mach number. The imposed adiabatic conditions ensure that the pressure is independent of the temperature. Since the velocities v and c are both slightly dependent on the pressure, then M turns out to be an effective surrogate for estimating the special attitude of the particle volume to change as a function of the pressure. In our investigations on practical ion drifts, the fluid can be assumed as being preserved from heat transfer at the boundaries and flowing with velocities v scarcely attaining some tens of meters per second. The former property for the fluid tacitly means that viscous dissipations and relevant heat conductions can be neglected; the second property is, for instance, typical of corona originated ions once they are injected into the extended drift region. Therefore, the condition of flow incompressibility, represented by the solenoidal nature of **v** and expressed in terms of div **v** = 0, will be adopted largely for the above class of problems.

3.9.3 Momentum Conservation

The relevant governing law is exactly represented by Navier's equation [see Eq. (3.26)], here reformulated taking into account Eq. (3.14).

Accordingly,

$$\frac{\partial \mathbf{v}}{\partial t} + \text{grad}\left(\frac{v^2}{2}\right) - \mathbf{v} \times \text{curl } \mathbf{v} = -\frac{1}{\rho_m} \text{grad } p - \frac{\mu}{\rho_m} \text{curl}(\text{curl } \mathbf{v}) + \mathbf{f}_m$$

(3.28)

As will be appreciated later with reference to subsonic and irrotational flows, the influence of the vorticity is immediately perceivable in Eq. (3.28). By the way, it is usual to replace the ratio μ/ρ_m, wherever it appears, with the symbol ν, which is representative of the so-called kinematic viscosity.

3.9.4 Total Kinetic Energy Conservation

Equation (3.27) reduces to

$$\frac{\partial(\rho_m e_t)}{\partial t} = 2\mu \, \text{div}(\boldsymbol{\mathcal{D}} \cdot \mathbf{v}) + \sigma_T \nabla^2 \, T - \text{div}(\rho_m e_t + p)\mathbf{v} + \rho_m(\mathbf{f}_m \cdot \mathbf{v}) \quad (3.29)$$

where

$$\boldsymbol{\mathcal{D}} \cdot \mathbf{v} = (\text{grad } \mathbf{v}) \cdot \mathbf{v} + \frac{1}{2}\mathbf{v} \times (\text{curl } \mathbf{v}) \tag{3.30}$$

∇^2 is the Laplace operator and, similarly to μ, σ_T has been treated as a constant.

3.10 INCOMPRESSIBLE AND IRROTATIONAL FLOWS

Imposing the condition curl $\mathbf{v} = 0$ in Eqs. (3.28)–(3.30) formally gives

$$\frac{\partial \mathbf{v}}{\partial t} + \text{grad}\left(\frac{v^2}{2}\right) = -\frac{1}{\rho_m} \text{grad } p + \mathbf{f}_m \tag{3.31}$$

$$\frac{\partial(\rho_m e_t)}{\partial t} = 2\mu \, \text{div}[(\text{grad } \mathbf{v}) \cdot \mathbf{v}] + \sigma_T \nabla^2 \, T - \text{div}(\rho_m e_t + p)\mathbf{v} + \rho_m(\mathbf{f}_m \cdot \mathbf{v})$$

(3.32)

for the momentum and energy conservation, respectively. Equation (3.31) exactly reproduces a familiar equation applied to ideal fluids, which, by definition, are inviscid ($\mu = \mu' = 0$) and irrotational. Note that div $\mathbf{S}$ drops out when the viscosity vanishes, so that the Navier–Stokes equations, Eq. (3.22), reduce to Eq. (3.31) if, additionally, curl $\mathbf{v} = 0$. In other words, the motion of an ideal fluid is undistinguishable from that of a viscous fluid whose flow is incompressible and irrotational. Bear in mind that viscosity, which is sensitive to pressure and temperature changes, has been tacitly assumed constant in the latter case in which a subsonic flow crosses a thermally insulated ambient.

3.11   DESCRIBING THE VELOCITY FIELD

3.11.1   Decomposition

Among the individual fields describing the flow field, special importance is ascribed to the continuous v-field which, in particular, is also admitted to be finite and to vanish at infinity. Therefore, Helmholtz's theorem [see Helmholtz (2009) and, for example, Morse and Feshbach (1953, p. 53)].

$$\mathbf{v}(\mathbf{r}, t) = - \operatorname{grad} \varphi_v(\mathbf{r}, t) + \operatorname{rot} \boldsymbol{\psi}_v(\mathbf{r}, t) \tag{3.33}$$

can be applied, where $\varphi_v(\mathbf{r}, t)$ and $\boldsymbol{\psi}_v(\mathbf{r}, t)$ denote the so-called Stokes scalar potential and vector potential, respectively. These quantities are not unique since any arbitrary scalar potential added to $\varphi_v(\mathbf{r}, t)$ and any arbitrary vector potential added to $\boldsymbol{\psi}_v(\mathbf{r}, t)$ cause no consequences on the equality expressed by Eq. (3.33). The negative sign assumed for the first component of $\mathbf{v}(\mathbf{r}, t)$ on the right-hand side of Eq. (3.33) is indicative of its orientation toward φ_v droping. Note that rot grad $\varphi_v = 0$ and div rot $\boldsymbol{\psi}_v = 0$, so that the first term of the above decomposition is the curl-less (irrotational) component, represented by the potential φ_v, of the total v-field, while the second term is the zero-divergence (solenoidal or isochoric) component represented by the potential $\boldsymbol{\psi}_v$. When the v-field is solenoidal, Eq. (3.33) reduces to $\mathbf{v} = \operatorname{rot} \boldsymbol{\psi}_v$; otherwise, Eq. (3.33) reduces to $\mathbf{v} = -\operatorname{grad} \varphi_v$ when the v-field is irrotational. From observing the general Eq. (3.33), it follows that div $\mathbf{v}(\mathbf{r}, t) = -\nabla^2 \varphi_v = 0$. Therefore, the part φ_v' of $\varphi_v = \varphi_v' + \varphi_v''$ representing the v-field with nonzero divergence obeys the Poisson equation $\nabla^2 \varphi_v' = \operatorname{div} \mathbf{v}$, the complementary part φ_v'' being harmonic. Likewise, resolving the Poisson equation $\nabla^2 \boldsymbol{\psi}_v' = -\operatorname{rot} \mathbf{v}$ gives the addend$\boldsymbol{\psi}_v'$ of the sum $\boldsymbol{\psi}_v = \boldsymbol{\psi}_v' + \boldsymbol{\psi}_v''$. Here, $\boldsymbol{\psi}_v'$ and $\boldsymbol{\psi}_v''$ respectively represent the nonzero rotational component of the v-field and the harmonic counterpart of $\boldsymbol{\psi}_v$.

3.11.2   The v-Field of Incompressible and Irrotational Flows

Setting rot $\mathbf{v} = \operatorname{div} \mathbf{v} = 0$ and using Eq. (3.33) yields

$$\operatorname{div} \mathbf{v} = \operatorname{div} \operatorname{grad} \varphi_v = \nabla^2 \varphi_v = 0 \tag{3.34}$$

Therefore, the v-field becomes Laplacian, which means that the scalar potential $\varphi_v = \varphi_v''$ ($\varphi_v' = 0$) is a harmonic function. Even the components of v and its modulus v are harmonic functions, since the partial derivatives of a harmonic function are harmonic as well. As regards the boundary of the domain under examination, the following statements hold:

- Maximum and minimum values of φ_v and v are only permitted on the boundary.
- The boundary conditions (which are an additional aspect of the problem specification) can be expressed in terms of surface distributions of φ_v or its normal derivative. Provided that nonarbitrary boundary conditions are assumed, the direct problem consisting of the evaluation of φ_v and its gradients in the internal points of the domain is indisputably well-posed. This is because the crucial condition of uniqueness is met in addition to the previously discussed existence and continuity ones.

A feature distinctive of incompressible flows is that the quantity $\int_V v^2 \, dV$ minimizes when the flow also becomes irrotational. This in mind, when spatiotemporal changes of ρ_m are also nonsignificant, even the total kinetic energy $\frac{1}{2} \int_V \rho_m v^2 \, dV = \frac{1}{2} \rho_m \int_V v^2 \, dV$ reduces to a minimum. This interesting behavior will be resumed in Section 3.12.2.

3.11.3 Some Practical Remarks and Anticipations

In principle, the notion of incompressible and irrotational flow introduces the same conceptual difficulties that arise for nonviscous fluids: On the boundary, the tangential component v_t of $\mathbf{v}$ and the velocity of the boundary itself are quite uncoupled quantities. By the way, in fluid mechanics the term boundary instead evokes a convoying rigid wall (lateral boundary) or an inner object, both "licked up" by the flow. Therefore, these objects analogically behave as nonconducting and impenetrable (permittivity theoretically tending to infinity) objects with reference to the electric field of a stationary electric current. However, $v_t > 0$ even if the boundary is at rest. Such an unphysical performance contradicts, in fact, the adherence (non-slip) condition expected to hold for viscous fluids (see Section 3.7). This implies that the notion of incompressible and irrotational flow of a viscous fluid should, contrary to physical experience, be unrealistic or should realistically join that of a fluid at rest. In the latter case, Eq. (3.22) gives, as expected, $\mathbf{f}_m = 0$ for grad $p = $ div $\mathbf{S} = 0$. The raised correspondence between fluid at rest and incompressible and irrotational flow results in theoretically assuming that the latter is steady, with zero force impressed. The apparent paradox, attributable to d'Alembert, of being $\mathbf{f}_m = 0$ for the flow under examination is in practice resolved, admitting that the field extension is not too vast and that the viscosity-related effects are confined in the boundary layers. Therefore, the steady flow conserving the momentum with zero body force can be safely applied to delimited domains, throughout (in a macroscale sense), because boundary layers are, however, reduced to pellicular structures. Of course, the momentum transfer by molecular collision must take place externally to the field domain. As will extensively be appreciated in Chapter 4, this approximate reading of a realistically subsonic and curl-less flow applies to ion flows. For such a class of problems, the ion-drift field could also be open, in the sense that the above-mentioned lateral wall could be nonexistent; after all, the flow remains rather concentrated. In these case studies, the boundary reduces, by elimination, to a pair of surfaces where the charges are injected or collected. Often, the injecting surface is an immaterial interface, closely surrounding an active electrode, positioned between the ionization (plasma) and drift regions; instead, the collecting surface is the impact one for ions drifting toward such a generally grounded electrode. Since the ion flow is subsonic and, as will be proved later, is guided by the pattern of an electrostatic field, then both the conditions pertaining to an incompressible and irrotational flow field are satisfied. Furthermore, inner objects, if any, are usually conductors, which implies that the fluid instead tends to flow in/out the body transversally to its surface. On the impact or emanation conducting surfaces an elemental electric charge can be envisioned as being neutralized ("subtracted" from an impacting ion) or transmitted to a neutral molecule (thus forming an ion). In that instant the impacting ion becomes, or, in the same order, the neutral is, macroscopically motionless, so that the usual notion of the stagnation point applies in spite of the presence of a nonzero electric field. In other words, all the electric fluxlines are interpreted as being stagnation streamlines according to fluid dynamics terminology. The theoretical d'Alembert paradox is safely circumvented, admitting a insignificant, even though not exactly null, body force of electrostatic origin acting within an ionized-flow carrying gap of finite dimensions. Rather, the significant amount of momentum transfer occurs in the restricted plasma region, or thereabout, where the charged supersonic constituents collide with the neutrals. Once pumped into the drift region, ion wind and ion drift are, in practice, mutually noninfluential flows, even forming a complex interpenetrative motion.

3.12 VARIATIONAL INTERPRETATION IN SHORT

3.12.1 Bernoulli's Equation for Incompressible and Irrotational Flows

Equation (3.28) can be reduced to the rearranged form

$$\text{grad}\left(\frac{v^2}{2} + \frac{p}{\rho_m}\right) = \mathbf{f}_m \qquad (3.35)$$

under steady conditions because ρ_m is constant for an incompressible flow and, additionally, rot $\mathbf{v} = 0$. Equation (3.35) tacitly advises that $\mathbf{f}_m$ can be represented by a potential energy per unit volume W such that $\mathbf{f}_m = -\text{grad}\, W / \rho_m$. Therefore, Eq. (3.35) can be written grad $U = 0$, where U is the constant energy density of the bulk motion equal to [see, for example, Hammond (1981) and Morse and Feshbah (1953, pp. 44–54)]

$$U = W_t + T_k \qquad (3.36)$$

with $W_t = W + p$. Note that U holds constant along any streamline and generally changes from a streamline to another. Equation (3.36) is termed Bernoulli's equation, subject to boundary conditions. Here, W_t expresses the total positional energy density due to external (through W) and internal (through p) forces; $T_k = \frac{1}{2}\rho_m v^2 = \rho_m e_k$ stands for the corresponding kinetic energy. Note how the existence of a scalar potential function W, whose gradient is proportional to the external force $\mathbf{f}_m$, results in the conservation of U, in which case the system (i.e., the flow) is termed *conservative*. Following a different theoretical route, it is permissible to claim energy conservation when the so-called Lagrange's function (or kinetic potential) $L = T_k - W_t$ is not an explicit function of time. The remaining two necessary and sufficient conditions for U to be conserved, namely, that W_t and T_k must respectively be independent on velocity and a quadratic form of the velocity, have clearly been satisfied in the present treatment. In this respect, bear in mind that a subsidiary feature acknowledged for subsonic (incompressible) flows is that p and v hold substantially uncorrelated (see Section 3.9.2); on the other hand, the quadratic form v^2 specifically required for T_k already appeared in Eq. (3.14) in the present treatment (see also Section 3.5.4). The flow is a continuum, which implies that the governing laws for the motion are differentiable and, therefore, invariant to coordinate transformations (see Section 3.3). Therefore, W_t and T_k are scalar invariants and, as such, suitable to be involved in a unique formulation describing the system. A set of similarly structured conservation laws can be derived from that original formulation. This subject will be touched upon in the following subsection where Lagrange's function plays an important role. In the light of these considerations, the existence of a general formulation of the kind represented by Eq. (3.6), which then specializes in the set of conservation laws obtained in Section 3.5, is fully accounted for. It should be remarked that Eq. (3.6) also incorporates, through $\mathbf{\Phi}$, dissipative terms connected to the viscosity. These quantities are generally accommodated in Lagrange's function involving total potential and kinetic energies. Accordingly, the original Noether's theorem, which acknowledges a one-to-one correspondence between conservation laws and symmetries of the action, can be claimed. The term *symmetry* is the mathematical substitute of *invariant* for a physical system subject to a class of transformations; action is defined as the integration of L over a given time interval [see Eq. (3.37)]. The theoretical difficulties that could at first arise in making the notions of conservation and dissipation somehow compatible are circumvented since the system has been assumed as being closed (thermally insulated). This ensures that energy transfers are allowed without loss of the total energy, which, rather, holds constant internally to the system itself. However, with special reference to the irrotational and incompressible flow of lossy fluids, the above remarks become unimportant since the flow behaves, as a matter of fact, as an ideal system.

3.12.2 Lagrange's Function

For a conservative flow, Hamilton's principle applies. Such a variational principle determining the equations of the fluid motion generally reads $\delta A = 0$ where

$$A = \int_{t_0}^{t_1} L \, dt \tag{3.37}$$

represent the action whose integrand is Lagrange's function. This function needs to be carefully specified for each case study. Classically, $L = T_k - W_t$ where, for the sake of opportunity, use has been made of the same notation of the previous subsection, even though the quantities now implied rather denote total energies of kinetic and potential nature. Setting $\delta A = 0$ advises that the action evaluated up to a given instant t_1 is a minimum for a flow subject to conservative forces and initial conditions (the latter expressed by the time instant t_0). Note that W_t could even vanish when an incompressible and irrotational flow is also steady, in which case $\mathrm{grad}(p/\rho_m + W/\rho_m) = \mathrm{grad}(W_t/\rho_m)$ needs first to be zeroed in Eq. (3.35) *before* performing further mathematical manipulations. Following this advice, Eqs. (3.35) and (3.36) reduce to $\mathrm{grad}\, T_k = 0$ (as expected, see below) and T_k, respectively, which implies that L can be unambiguously represented by T_k alone. Note that responsibility for the Laplacian nature of the velocity field (see discussion in Section 3.11.2) is, in accord with a variational reading, the vanishing contribution of W_t associated with a subsonic and irrotational flow subject to zero conservative body forces. In fact, the Laplace equation governing the $\mathbf{v}$-field is exactly represented by the Euler equation when $L = T_k$. Bear in mind that *a priori* extending the condition $W_t = 0$—distinctive of a fluid at rest, even deprived of gravitational forces—to moving fluids has implied, according to the arguments raised in Appendix 3.A, that a latent material quality, conceptually independent of the spatial characteristics, time instant t_0 in which the **v**-field is established, and duration $(t_1 - t_0)$ of the relevant action, does exist. Otherwise, setting $L = T_k$, instead of $T_k - W_t$, only because the nonzero constant quantity W_t has been deliberately neglected, appears to be an arbitrary assumption. Accordingly, care should be taken in preliminarily structuring the functions being differentiated in space—this is the case for Eq. (3.35), which specializes under a form of grad $T_k = 0$—and time in conformity to the fluid matter's attributes indicated above. It has been proved that the raised formal incongruence about the value assigned to W_t is resolved by reasoning in terms of the conserved (field-history-independent) material property applied to the fluid. This crucial notion will be reintroduced and corroborated in Chapter 4, Section 4.3.3, as regards the limits of application even affecting Gauss's law.

Last, it is worth noting that for an incompressible and irrotational flow, the action

$$A = \int_{t_0}^{t_1} T_k \, dt \tag{3.38}$$

is stationary and its minimization corresponds to the minimization of the one-dimensional integral $\int_L T_k dL$ performed along any given streamline of length L in the Laplacian **v**-field. This additionally implies the minimization of the volume integral $\int_V v^2 \, dV$ extended all over the field domain, as already considered at the end of Section 3.11.2. Note that if v holds constant along a streamline, then $\int_L T_k dL = T_k L = $ const and, in turn, $vL = $ const for each streamline. As will be carefully shown in Chapter 4, the linear proportionality between ion drift velocity **v** and electric field **E** in the ion-flow field will imply that the product $E{\cdot}L$ holds constant along the ion trajectory guided by the streamline of length L. The above constant value identifies the voltage applied to the electrode system (boundary condition), namely, the common potential difference applied to each trajectory.

Parenthetically, a constant value of the modulus of **v** along the streamline allows Eq. (3.38) to become

$$A = T_k \int_L dL \tag{3.39}$$

Equation (3.39) expresses the principle of stationary (or least) action, extended over all the trajectory of finite length, of the particle in the **v**-field (or in the **E**-field). This result is expected since it has been ascertained that Maxwell's equations and, in particular, electrostatic fields can be derived from the above principle. Note that the prerequisite for a scalar function φ_v to be harmonic, hence, for the **v**-field to be Laplacian (see Section 3.11.2), is that φ_v at any given point is equal to the average value of φ_v in the neighborhood. This property is consistent with the notion of least action when the integral A is rather performed over short arcs of the entire L or when the gradient of φ_v is constant along a fluxline arc of arbitrary length. As a consequence, Laplacian streamlines can be rigorously interpreted as paths of shortest time τ and, simultaneously, of shortest length L when the particle travels with constant velocity in the space of the generalized coordinates. Therefore, it could be stated, in anticipation of the subject of Section 4.3.6, that the often cited time-of-flight of drifting ions is as short as it can be, thereby allowing us to assume the same notation τ previously introduced for minimized times.

APPENDIX 3.A

According to Eq. (3.10), the mass conservation stated in the co-moving reference frame can be rewritten

$$\frac{Dn}{Dt} = -n \operatorname{div} \mathbf{v} \qquad (3.A.1)$$

after remembering that $\rho_m = nm_0$, the quantity m_0 being a constant (see Section 3.3). For a moving particle composed of the same number $n\,\delta V$ of molecules, the substantial derivative $D(n\,\delta V)/Dt$ is identically equal to zero 0, so that

$$\frac{Dn}{Dt} = -\frac{n}{\delta V}\frac{D(\delta V)}{Dt} \qquad (3.A.2)$$

easily derives. Therefore

$$\frac{D(\delta V)}{Dt} = \delta V \operatorname{div} \mathbf{v} \qquad (3.A.3)$$

is the result of a simple comparison between Eqs. (3.A.1) and (3.A.2). Provided that the fluid behaves as a perfect gas subject to reversible and isentropic processes, the equation

$$\frac{3}{2}nK_B\frac{DT}{Dt} + \frac{p}{\delta V}\frac{D(\delta V)}{Dt} = 0 \qquad (3.A.4)$$

legitimately applies (K_B represents Boltzmann's constant). Equation (3.A.4) shows how the first law of thermodynamics is implied when the fluid particle moves. Eq. (3.A.3) allows Eq. (3.A.4) to alternatively read

$$\frac{3}{2}nK_B \frac{DT}{Dt} + p \operatorname{div} \mathbf{v} = 0 \tag{3.A.5}$$

where

$$nK_B = \frac{p}{T} \tag{3.A.6}$$

represents the basic equation of state for perfect gases expressed in functions of n. It should be remembered that distinctive of an incompressible flow are the identities div $\mathbf{v} = 0$ and $n = \operatorname{const}$ with respective regard to Eqs. (3.A.5) and (3.A.6). As a combinative result, the pair of ambient parameters T and p are verified to hold unchanged during the motion, so that

$$\frac{DT}{Dt} = \frac{Dp}{Dt} = 0 \tag{3.A.7}$$

can be set formally. On the other hand, the thermodynamic equation of state governing the density ρ_m can be expressed as

$$\frac{1}{\rho_m} \frac{D\rho_m}{Dt} = \alpha \frac{Dp}{Dt} - \beta \frac{DT}{Dt} \tag{3.A.8}$$

where

$$\alpha(p, T) = \rho_m^{-1} \left[\frac{\partial \rho_m}{\partial p} \right]_T,$$
$$\beta(p, T) = -\rho_m^{-1} \left[\frac{\partial \rho_m}{\partial T} \right]_p \tag{3.A.9}$$

represent thermodynamic quantities, strictly pertaining to the fluid materiality, referred to, in the above order, as isothermal compressibility and bulk expansion coefficients [see Panton (1984)]. Instead,

$$\frac{Dp}{Dt} = \frac{\partial p}{\partial t} + (\operatorname{grad} p) \cdot \mathbf{v},$$
$$\frac{DT}{Dt} = \frac{\partial T}{\partial t} + (\operatorname{grad} T) \cdot \mathbf{v} \tag{3.A.10}$$

are the material derivatives strictly pertaining to fluid dynamics.

For a medium at rest, $\partial p / \partial t = \partial T / \partial t = \mathbf{v} = 0$, which implies that both Dp/Dt and DT/Dt drop to zero. But the identity $(1/\rho_m)(D\rho_m/Dt) = 0$ so given in Eq. (3.A.8) also derives in the dynamic case of an incompressible flow, for which, as previously verified, the identities expressed by Eq. (3.A.7) hold. The implication is that a fluid at rest and a subsonic flow are indistinguishable as regards ρ_m performances along any given streamline. Such a property can be legitimately interpreted as a perpetual character of the fluid, in the sense that when a fluid turns on from its initial immobility, ρ_m remains constant everywhere in the domain under examination, provided that ρ_m is constant at the boundary and an incompressible excitation field is established at a given time instant. In some cases, ρ_m holds rigidly constant only along any single streamline, in the sense that ρ_m is allowed to change from a streamline to another, subject to boundary conditions. This is the case for the present ion drift model derived from corona activity, where ions are injected according to specified distribution laws at the boundary and the involved **v**- and **E**-fields in the drift region are irrotational rather than solenoidal. As a consequence, the equivalence discussed in Section 3.9 between incompressible viscous flow and nonviscous flow, when both are irrotational, becomes important. This equivalence is inherently independent of the states of motion or at rest for the fluids and, therefore, it is permissible to admit that a viscous fluid at rest conserves, even though in a latent state the aptitude for incompressibility, along with a nonviscous (inviscid) fluid, at rest conserves its latent insensitivity to viscous effects. However, a rigorous argument involving Eq. (3.A.8) can be adopted to vindicate the conceptual transfer to the fluid material of those physical features that are seemingly related to the excitation field and its history. To this end, note that the pair of Eqs. (3.A.9) can be profitably rewritten as

$$\alpha(p, T) = \rho_m^{-1} \left[\frac{\partial \rho_m / \partial t}{\partial p / \partial t} \right]_T, \qquad \beta(p, T) = -\rho_m^{-1} \left[\frac{\partial \rho_m / \partial t}{\partial T / \partial t} \right]_p \qquad (3.A.11)$$

so that $\partial \rho_m / \partial t$ can be replaced, according to Eq. (3.10), with $-\rho_m \, \text{div} \, \mathbf{v}$ while the remaining pair of time derivatives $\partial p / \partial t$ and $\partial T / \partial t$ can respectively be replaced, according to the pair of Eqs. (3.A.10), with $Dp/Dt - (\text{grad} \, p) \cdot \mathbf{v}$ and $DT/Dt - (\text{grad} \, T) \cdot \mathbf{v}$. Accordingly,

$$\alpha = -\frac{\text{div} \, \mathbf{v}}{\frac{Dp}{Dt} - (\text{grad} \, p) \cdot \mathbf{v}}, \qquad \beta = \frac{\text{div} \, \mathbf{v}}{\frac{DT}{Dt} - (\text{grad} \, T) \cdot \mathbf{v}} \qquad (3.A.12)$$

are permissible reformulations of the starting Eqs. (3.A.9), which at last become

$$\alpha = \frac{\text{div} \, \mathbf{v}}{\frac{2p \, \text{div} \, \mathbf{v}}{3} + (\text{grad} \, p) \cdot \mathbf{v}}, \qquad \beta = -\frac{\text{div} \, \mathbf{v}}{\frac{2T \, \text{div} \, \mathbf{v}}{3} + (\text{grad} \, T) \cdot \mathbf{v}} \qquad (3.A.13)$$

In Eqs. (3.A.13), the equalities $DT/Dt = -2T \operatorname{div} \mathbf{v}/3$ and $Dp/Dt = (p/T)(DT/Dt) = -2p \operatorname{div} \mathbf{v}/3$ have been adopted by virtue of Eqs. (3.A.5) and (3.A.6). The merit ascribed to the set of Eqs. (3.A.13) consists of showing unsophisticated and eloquent relationships between the material properties α and β and fluid-dynamic quantities, the latter involved in the set of Eqs. (3.A.10). In fact, Eqs. (3.A.13) reduce to

$$\alpha = \frac{3}{2p}, \qquad \beta = -\frac{3}{2T} \qquad\qquad (3.A.14)$$

which, restrictively speaking, hold constant along any given fluxline without any regard as to the status of at rest or subsonic motion (gravitation neglected), namely, of the excitation field's establishment or suppression. Often, in a thermally insulated ambient crossed by an incompressible flow, p and n and, therefore, p and T can be assumed to be constant everywhere in the field domain, which implies that the material properties represented by α and β are independent of the excitation field and, therefore, of its establishment or removal. This field-history-independent material performance is intimately connected to the crucial condition $\operatorname{div} \mathbf{v} = 0$ that is formally applicable to a fluid at rest (trivial case) or flowing subsonically (nontrivial case). As a consequence, the above condition is somehow also evocative of a perpetual fluid property. This reasoning holds its validity even along unidimensional fields, namely, when α and β are permitted to change from a streamline to another because, say, of the topological structure of a discontinued (multichanneled) $\mathbf{v}$-field. This is the scenario pertaining to ion swarms under the action of an applied electric field (see Chapter 4). There, the in-tandem identity $\operatorname{div} \mathbf{E} = 0$ due to the relationship $\mathbf{v} = k \mathbf{E}$ involving the constant mobility k, will be simultaneously adopted in going on in the formalization of the drift model (see Section 3.12.2). Setting $\operatorname{div} \mathbf{E} = 0$ and likewise interpreting such an identity as evocative of a fluid property, independent of such a medium to be at rest or in motion, will be consequential with special reference to a revised application of Gauss's law (see, in particular, Section 4.3.3).

CHAPTER 4

ELECTROHYDRODYNAMICS OF UNIPOLAR ION FLOWS

4.1 INTRODUCTION

The contents of previous chapters, with special reference to Chapters 2 and 3 (see, in particular, Section 3.1), provide the indispensable informatory background necessary before introducing the coupled model. According to this theoretical resource, that fundamental subject, which is the electric conduction field in a material substance, can no longer be interpreted as being incidentally analogous to a fluid flow; rather, substantial arguments will be provided to appreciate how the two kinds of fields result, as a matter of fact, indistinguishable in a unified and discernible phenomenology. Therefore, the ion flow identifies with a subsonic ion drift guided by the pattern of an electric field, and, as such, it can only assume a filamentary space structure. This issue, carefully described in Section 4.4, is what characterizes the present chapter as a whole. As will be appreciated later, such an appealing model on one hand pushes aside previously questionable theories, restores validity to debated hypotheses, and enables special account to be taken of some previously claimed analogies of recurring interest. On the other hand, it offers powerful insights into those injection mechanisms that lead one to determine additional key conditions at the boundary. The ultimate intent is that of making computational problems subject to such conditions finally reliable, other than well-posed. Some ancillary sections will progressively introduce

Filamentary Ion Flow: Theory and Experiments, First Edition. Edited by Francesco Lattarulo and Vitantonio Amoruso.
© 2014 by The Institute of Electrical and Electronics Engineers, Inc. Published by 2014 John Wiley & Sons, Inc.

the reader into the fascinating multichanneled scenario of moving ions. In doing so, it will be worth addressing a number of preparatory arguments, whose inherent novelty makes them of prominent and widespread individual interest. To give just some examples, this has been the opportunity to talk about the combined mass-charge notion (Section 4.2) and the specialized conservation laws for charge, mass, momentum, and energy (Sections 4.3.1 and 4.3.6), as well as about Gauss's law (Section 4.3.3), ion mobility (Section 4.3.7), and Deutsch's hypothesis (Section 4.3.8). According to the central Section 4.4, other notable subjects require to be revisited; this is the case for the ion-drift equation and ion wind generation and propagation, also treated elsewhere (Sections 4.5.1 and 4.5.3, respectively).

4.2   REDUCED MASS-CHARGE

Before going on with the treatment, it is important to remember what has been stated earlier (Chapter 3, Sections 3.1 and 3.3) as regards the fixed identity of each fluid particle forming the flow. The molecules contained in each elemental particle must hold the same (so as to form a closed subsystem) during the motion in order for the basic fluid dynamic theory illustrated in Chapter 3 to be applied. This condition is, for example, met when the flow is determined by an electric field applied to charge-bearing molecules in free space, but not when these give rise to an electrical conduction. In the latter case, in fact, a nearly unchanging number of motionless neutrals, positioned in a given time instant inside/on the moving particle, are continuously replaced during the motion. On the other hand, the molecules permanently residing in/ on the moving particle identify with the drifting ions. Therefore, the charged molecules capable of moving are only a portion of the total fluid mass filling the particle, the complementary part being composed of motionless neutrals. Indeed, the described objection is in no way daunting, in the sense that the fluid dynamic theory can be successfully recovered by *a priori* invoking the two-body model. The term *body* is adopted here in a microscale sense, thus applied to an individual ion and its neutral counterpart. As a substantial result, a true conduction in gases—this is the case for ion drift—appears theoretically equivalent to a convection of charged molecules as soon as the interpenetrative two-body fluid system made of moving and at-rest masses is virtually replaced by a collectively moving one-body mass of fluid. This model is fully consistent with the notion of a particle having fixed identity and, therefore, suitable for application to a coupled ion-flow formulation. With reference to the total number density n of the gas, only the fractional amount $n_c < n$ of solute ions of mass m_c is really capable of drifting, under the influence of an electric field, through a buffer medium composed by neutrals of mass M.

Using the two-body model [see, for example, Landau and Lifshitz (1971)], even the neutrals represented by the remaining $n - n_c$ particles per unit volume are envisioned to ride with the same velocity $\mathbf{v} = k\,\mathbf{E}$ of the drifting n_c ions. In the above equality, k denotes ion mobility, assumed constant for the reasons that will be explained extensively in the following subsection. The unphysical inclusion of neutrals in the flow model is made permissible because the totality of n gas particles per unit volume virtually acquires the same mass m and charge q, both to be specified. In fact, m and q are adjusted constants, here referred to as reduced mass $(1/M + 1/m_c)^{-1}$ of the ion-neutral collision pair and reduced charge $q_c n_c / n$. In particular, m can be assumed constant without incurring an appreciable error, provided that m_c and M are comparable masses. This seems to be the case for the present class of problems in which m_c is substantially determined by a dominant species of light ions. Radioactive and corona sources especially produce light ions whose charge density is prone to exceed the sum of the remaining component species and, hence, approaches the overall value. Such a prevailing class of ions are formed by hydrated particles that tend to conserve molecular size (order of 1 nm), mass m_c, and charge q_c. Note that the formation process for the above conservation to be accomplished is rather quick (hardly exceeding 1 μs). This means that mobility changes due to collisions with the buffer gas during usual lifetimes (time of flight) can be neglected—a practical estimation corroborated by large amounts of experimental data in the atmosphere. These show narrow mobility spectra for both polarities, even in the case of remotely monitored ion flows with lifetimes estimated at several seconds [see, for example, Johnson and Zaffanella (1983)]. The averaged absolute values ranging between 1 and 2 cm^2/(Vs) are distinctive of air under the usual conditions [see also, for example, Misakian, Anderson, and Laug (1989)], with the stipulation of tacitly admitting larger absolute values for negative ions because of electron shares in the overall flow. The residual contribution of drifting electrons is of course due to the rather weak electronegative character of air. Returning to the above model, q is constant by definition because the charge per unit volume $\rho_c = q_c n_c$, physically related to the n_c solute ions, is distributed in the model over all the n gas particles. Accordingly, $q = q_c n_c / n < q_c$, with $n_c/n < 1$, and $\rho = qn = \rho_c = \text{const}$. Consider that the identity between the charge densities ρ_c and ρ is strictly connected to the discussed equivalence between convective and conductive flows, even though ρ_c is rather akin to the notion of convection and ρ to that of conduction.

Note further that a possible formulation for the ion mobility is, according to Revercomb and Mason (1975),

$$k = \frac{\zeta q}{m} \tag{4.1}$$

where

$$\zeta = \frac{3 \left[\dfrac{2\pi m}{K_b T} \right]^{1/2}}{16 n_c \left(1 - \dfrac{n_c}{n} \right) \Omega_0}$$

is an ancillary parameter, thus also depending on Boltzmann's constant K_b, drift gas temperature T, and ion's collision cross-section Ω_0 [see also Mason and McDaniel (1988) and Eiceman and Karpas (2005)]. Note that ζ has the dimension of time and can assume the physical meaning of the collision period. The property of ion mobility to be nearly constant for a gas is connected to the slight influence of the **E**-field on ζ and m (the so-called low-field assumption). Instead, the discussed dependence of k on polarity, due to a modest influence of concurring free electrons to the overall negative drift, can be taken into account by simply increasing the absolute value of q, with m left unchanged. In light of the present coupled model, more details regarding ion mobility can be found in Section 4.3.7.

4.3   UNIFIED GOVERNING LAWS

4.3.1   Mass-Charge Conservation Law

This subsection shows how a typical example of analogy, pertaining to the present study, is in reality indicative of a prominent physical coupling. In other words, insights here involve the interpretation of the analogous charge and mass conservations laws as different ways to express a unique charge-mass conservation law. This applies to the ion-neutral complex according to a unified model. Let us consider the following pair of continuity differential equations extensively pointed out in basic textbooks as analogical formulas:

$$\frac{\partial \rho}{\partial t} + \mathrm{div}(\rho \mathbf{v}) = 0 \tag{4.2}$$

$$\frac{\partial \rho_m}{\partial t} + \mathrm{div}(\rho_m \mathbf{v}) = 0 \tag{4.3}$$

Equations (4.2) and (4.3) respectively govern charge and mass conservation [Eq. (4.3) is the same as Eq. (3.8)]. The product $\rho \mathbf{v}$ is the current density $\mathbf{J} = k\rho\mathbf{E}$ due to ions drifting with reduced-mass density ρ_m and reduced-charge density ρ (the notation for the reduced quantities has been left unchanged with respect to the unreduced counterparts without creating confusion).

Parenthetically, consider that the current density is also affected by a diffusional mechanism, even though it is assumed to be negligible during the ion drift. This is because diffusion is postulated as being practically confined in a peripheral layer (henceforth referred to as the diffusion layer) of the ionization region. Such an overall region is intimately attached to the active conductor and occupies a restricted volume in comparison to the outwardly extended drift space. Under some specialized circumstances, a significant portion of the diffusion layer happens to trespass on and lick up the surrounding nonactive zones of the same conductor. During this expansive phase, the thickness of the above layer and, hence, the shape of the ionization region change until a stable outer geometry is attained. As will be appreciated later, an important consequence is that space distribution of the ions prone to be definitively repelled into the drift region are remarkably influenced (Appendix 4.A). Based on the above description, the strategy adopted here consists of neglecting ion diffusion during the drift [see, for example, Sigmond (1978)], with the stipulation that the diffusional phenomenon has been estimated in advance, namely, through a supplementary investigation into the outer morphology of the ionized region. Useful advice on this concern can be found in Appendix 4.B because the peripheral diffusion layer can be significantly involved in the course of shaping (cross-references are somewhere disseminated in the present and next chapter).

Our task is now to find a still overlooked physical relationship to which the striking analogy between Eqs. (4.2) and (4.3) must be subordinated. Put another way, Eq. (4.2) could be directly extracted from Eq. (4.3) (or vice versa), provided that both sides of the latter formula are multiplied by a suitable constant quantity having physical meaning. Indeed, the prerequisite for this constant to be identified has been adequately discussed before: The molecules in the volume and on the boundary of an elemental volume of fluid (the particle) must be the same during the motion. Accordingly, it is exactly the constant ratio q/m embodied in Eq. (4.1) that is asked for playing the key role of deriving Eq. (4.2) from Eq. (4.3), in that identified as a charge density per unit mass, $q/m = qn/(mn) = \rho_0$ clearly represents an invariant of the system. Therefore, multiplying both terms of Eq. (4.3) by the constant ρ_0 gives Eq. (4.2) because $\rho = \rho_0 \rho_m$. This simple but attractive result has been made permissible, just theoretically making (as illustrated before), a true conduction of ions undistinguishable from an untrue convection subject to established conditions. These impose that all the gas molecules virtually acquire reduced mass and charge in the original drift region where they are set in a collisionless motion with impressed velocity $\mathbf{v}$ (Appendix 4.B). On the substance, it is permissible to admit the following:

- The physically meaningful linear laws $\rho = \rho_0 \rho_m$ and $\mathbf{v} = k\mathbf{E}$ behave as coupling formulas involving scalar and vector quantities. These are pertaining to the traditional realms of electrostatics (ρ and $\mathbf{E}$) and fluid dynamics (ρ_m and $\mathbf{v}$). As an accessory observation, the ρ- and ρ_m-fields are coincident up to the constant ρ_0; therefore, they form a pair of inherently analogical fields. The same considerations can be repeated for the $\mathbf{v}$- and $\mathbf{E}$-fields, which are coincident up to the constant k.

- The coupled densities ρ and ρ_m are directly responsible for the apparent analogy between Eqs. (4.2) and (4.3).
- The formal identity between drift and convection currents, both expressed through the common formula for the density $\mathbf{J} = \rho\mathbf{v}$, is indicative of a mutual equivalence appreciable when the pair of reduced quantities m and ρ is introduced in combination with the coupling relationship $\mathbf{v} = k\,\mathbf{E}$, the latter traditionally pertaining to conduction.

4.3.2 Fluid Reaction to Excitation Electromagnetic Fields

For an exhaustive description of this class of problems especially informed with engineering paradigms, the set of governing electromagnetic equations summarized below require a preliminary introduction:

$$\text{curl }\mathbf{E} = \frac{-\partial\mathbf{B}}{\partial t} \tag{4.4}$$

$$\text{curl }\mathbf{H} = \mathbf{J} + \frac{\partial\mathbf{D}}{\partial t} \tag{4.5}$$

$$\text{div }\mathbf{B} = 0 \tag{4.6}$$

$$\mathbf{D} = \varepsilon\mathbf{E} \tag{4.7}$$

$$\mathbf{B} = \mu\mathbf{H} \tag{4.8}$$

$$\mathbf{f_m} = \mathbf{E}\,\text{div }\mathbf{D} + \mathbf{J} \times \mathbf{B} - \frac{1}{2}(\mathbf{E}^2\,\text{grad }\varepsilon - \mathbf{H}^2\,\text{grad }\mu)$$
$$+ \frac{(\varepsilon_r\mu_r - 1)}{c_2}\partial(\mathbf{E} \times \mathbf{H})/\partial t \tag{4.9}$$

According to a usual notation for vector quantities, $\mathbf{H}$, $\mathbf{B}$, and $\mathbf{D}$ denote, respectively, magnetic field, magnetic flux density, and electric displacement; the scalar parameters ε and μ represent the medium's permittivity and permeability, respectively (whose dimensionless relative values are subscribed by r); c denotes speed of light in free space. Note the following:

- Equations (4.4) and (4.5) form the pair of *stricto sensu* Maxwell's equations.
- Equation (4.6) expresses the solenoidal nature of $\mathbf{B}$.

- Equations (4.7) and (4.8) are subsidiary relations required for the electromagnetic characterization of the medium.
- Equation (4.9) is an exhaustive formulation for the force exerted on a unit volume of isotropic medium.

Equation (4.9) is of paramount importance in estimating the real performances of a material filler subject to the influence of an electromagnetic field. In other words, it is the mechanical reaction of the material which is responsible for the true excitation features of the applied field. Before going on in this investigation, substantially based on simplifying Eq. (4.9), a conceptual difficulty needs to be dissipated directly. Just now, the reader is invited to pay attention to the first addend on the right-hand side of Eq. (4.9), where div **D** is directly given after performing some mathematical manipulations. Incidentally, the addend under examination replaces, according to tabulated vector identities, the difference $\text{div}(\mathbf{DE}) - (\mathbf{D}\cdot\text{grad})\mathbf{E}$ involved in the proof leading to Eq. (4.9). In particular, $\text{div}(\mathbf{DE})$ is the divergence of a tensor whose physical meaning becomes discernible if reference is made instead to the associated integral form

$$\int_V \text{div}(\mathbf{DE})dV = \int_A (\mathbf{D}\cdot\mathbf{n})\mathbf{E}dA \qquad (4.10)$$

As usual, **n** represents the outwardly directed unit vector on the surface A bounding the volume V of the domain under examination (say, a drifting region). The quantity div **D** at last appearing in Eq. (4.9) is deliberately not replaced, according to Gauss's law div $\mathbf{D} = \rho$, with the charge density ρ. The subtle motivation accounting for such an appropriate restraint requires careful examination since preference is customarily expressed, because of Gauss's law, for ρ in substitution of div **D** [see Stratton (2007) as an influential example]. Instead, this common practice is in the present case strongly warned, as will be illustrated in the following subsection.

4.3.3 Invalid Application of Gauss's Law: A Pertaining Example

The law of the conservation of charge, before expressed according to Eq. (4.2), assumes the integral form

$$\int_S \mathbf{J}\cdot\mathbf{n}\,da = -\frac{d}{dt}\int_V \rho\,dV \qquad (4.11)$$

Equation (4.11) formalizes the axiomatic concept, applicable to our problems in which chemical reactions represented by the charge's source or sink are absent, that charge may not be created or destroyed. The surface S on which the normal unit vector $\mathbf{n}$ points outwards encloses a control volume V (fixed in time and containing a certain amount of charge at the instant t). Taking the time derivative in Eq. (4.11) inside the integral, along with applying the divergence theorem and converting the result in differential form, at last gives Eq. (4.2). Alternatively, applying the divergence theorem to Ampères's circuital law [see Eq. (4.5)] and remembering that μ div $\mathbf{H} = 0$ gives

$$\text{div } \mathbf{J} + \frac{\partial \text{div}(\varepsilon \mathbf{E})}{\partial t} = 0 \qquad (4.12)$$

Rearranging Eq. (4.12) and Gauss's law differential form $\text{div}(\varepsilon \mathbf{E}) = \rho$, it again follows Eq. (4.2) since $\mathbf{J} = k\,\mathbf{v}$. It is worth considering, however, that the first way to get the continuity equation is indeed more general than the second for recourse as Gauss's law has been circumvented because of starting from Eq. (4.11). This observation is of paramount importance in order to avoid shortcomings and misinterpretations that could arise in the treatment of some "moving material" fields. In fact, comparing the primary Eqs. (4.2) and (4.12) yields

$$\frac{\partial [\text{div}(\varepsilon \mathbf{E}) - \rho]}{\partial t} = 0 \qquad (4.13)$$

The expression under the sign of the partial derivative becomes zero if admission is made that the field currently in existence was first established at some time in its history [see, again, Stratton (2007)]. Based on this observation, Gauss's law is the result of straightforward mathematical manipulations involving Eq. (4.13). It is immediately apparent that Gauss's law, expressed in its full formulation, is a derived one and, in that regard, its general validity strictly depends on whether the above-mentioned manipulations are always permissible or not. In the affirmative case, the stationary $\mathbf{E}$-field of steady currents invariably will be identical to the electrostatic field of correspondingly distributed charges at rest. So far, extensive evidence has been given elsewhere that the field calculation in the presence of unipolar ion drift tacitly admits, through Poisson's equation, the validity of Gauss's law in its complete form with ρ different from zero. Unfortunately, this admission is questionable since Gauss's law, applicable in its full formulation to massless or motionless material electric fields without exception, cannot be arbitrarily transferred to electric fields acting on fluid matter susceptible to motion. This is especially true when flows of charged gases are under examination, in which case Gauss's law with ρ different from zero is formally inconsistent with the incompressible nature of a fluid flowing with subsonic velocity. In fact, even though a gaseous fluid is compressible, subsonic dynamics results in flow incompressibility and amounts to formally set div $\mathbf{v} = 0$ (zero rate of expansion/contraction for the fluid particle) in conjunction with div $\mathbf{E} = 0$, $\mathbf{v}$ being expressed as $k\mathbf{E}$. Therefore, *a priori* introducing the identity div $\mathbf{v} = 0$ is only a mathematical device to mean an inherently perpetual fluid property. The latter is independent of whether it manifests in the form of solenoidal $\mathbf{v}$- and $\mathbf{E}$-fields at some time in the finite past. Differently speaking, the physical inaptitude for a fluid particle to expand/contract is irrespective of the fluid's excitation history, thus of being the fluid mass in motion or at rest. With reference to the discussion in Section 3.9.1, the spatiotemporal invariance of ρ_m therefore corresponds perfectly to the invariance of ρ so that $\mathbf{v}$ and ρ_m and, simultaneously, $\mathbf{E}$ and ρ are mutually independent. As a consequence, unsteady conditions for the traveling ions are interdicted even in the crucial instants in which the excitation $\mathbf{E}$-field is established or removed.

Such observations are perfectly consistent with the variational approach discussed in Section 3.12.2 and Appendix 3.A. Incidentally, it is worth noting that the above discussed invariance of v during the ion drift was also obtained by Kaune, Gillis, and Weigel (1983) by a different variational procedure based on minimizing the total flight time after neglecting any hypothetical relationship between $\mathbf{E}$ and ρ. The involved investigators felt doubtful in fully vindicating the actual value of the result because the adopted procedure is, in fact, made practicable by waiving Gauss's and continuity laws. In hindsight, the present analysis proves how the assumption of incompressibility, legitimately applicable to corona-originated ion flows, exactly establishes defensibility to the above-mentioned issue. As it turns out now, this revaluation is attributable to the zero rate of particle expansion under subsonic dynamics (div $\mathbf{v} =$ div $\mathbf{E} = 0$) which causes $\mathbf{E}$ versus ρ relationships, exactly represented by Gauss's and continuity laws, to break up.

On the other hand, the above considerations are inapplicable, say, to the prosaic case of a solid dielectric since the material reaction, surrogated by the permittivity ε, is in fact uninfluential for the patterns of both the excitation $\mathbf{E}$-field and the reaction $\mathbf{D}$-field, the latter only being subject to transient polarization processes that start as the exogenous $\mathbf{E}$-field establishes. Therefore, Gauss's law can be legitimately extracted from Eq. (4.13) by integration since $\mathbf{D} = \varepsilon\mathbf{E}$ was zero before applying the $\mathbf{E}$-field and will return to vanish upon removal of the same $\mathbf{E}$-field.

4.3.4 Laplacian Field and Boundary Conditions

As a formal consequence of the above reasoning, the terms containing div $\mathbf{E}$ under the sign of differentiation in Eqs. (4.12) and (4.13) drop out *before* performing further mathematical techniques involving time derivatives. Accordingly, the **J**- and **E**-fields are bound to become solenoidal and Laplacian, respectively. In other words, the field source is expected to reside externally to, or at the boundary of, the drift region because unsteady dynamics giving rise to charge accumulation is interdicted within ($\partial\rho/\partial t = 0$). Incidentally, this reason explains why initial conditions, generally accompanying the boundary ones, are ignored in the present context. It is worth noting that the Laplacian nature assumed by the electrostatic field of such a fluid matter in the interelectrode spacing is not to be intended as connected to a space-charge-free, but rather to a space-charge-accumulation-free, condition since the EHD coupling is just a matter of field matching. In other words, merely setting $\mathbf{v} = k\mathbf{E}$ must warn us that a complex material reaction is admitted: The originally decoupled solenoidal **v**-field and irrotational **E**-field are, respectively, forced to dually become irrotational and solenoidal. Overall, the above pair of fields are clearly permitted to join a unique Laplacian structure in spite of the presence of potentially perturbing unipolar space charge. This appealing result derives from a combinative interaction mechanism, involving original fluid dynamic and electromagnetic properties, capable of unambiguously determining coupled performances for the ion flow. Because they are both Laplacian, the above pair of joined fields are only subject to boundary conditions demanding for the just-generated ions explicit formulation of the injection law into the drift region. Unfortunately, the boundary under examination identifies with an imaginary interface between the ionization and drifting regions, so that establishing the above law resulted in an arbitrary exercise. A subsidiary problem consists of appropriately featuring the configuration of such an interface that, in general, is presumed to differ from the shape of the active conductor. Overall, assigning appropriately the actual geometry of and, consequently, current distribution at the interface has been the hard prerequisite for a successful computational approach. The multichanneled ion-flow model, carefully discussed hereinafter (Section 4.4), will give insights into the specification of such a challenging problem. However, denoting the applied voltage by V and denoting the length of a given streamline by L will give $E = V/L$ representing the magnitude of the constant electric field along that material line. This implicitly means that the velocity imparted to the just-injected ions is normal to the boundary surface and that the voltage drop in the untreated ionization region is comparatively negligible.

4.3.5   Vanishing Body Force of Electrical Nature

At the first stage of a permissible simplification, Eq. (4.9) reduces to

$$\mathbf{f}_m = \varepsilon_0 \mathbf{E} \operatorname{div} \mathbf{E} + \mu_0 \mathbf{J} \times \mathbf{H} \tag{4.14}$$

thus becoming a possible formulation of Lorentz's force law in free space. In fact, the equalities $\varepsilon_r = \mu_r = 1$ and grad $\varepsilon_0 =$ grad $\mu_0 = 0$ also apply to a gaseous medium (subscript 0 indicates free-space parameters). This means that the total force is deprived of the electromagnetic component embodied in the original formula. Additionally, an educated guess applied to the present class of investigations consists of setting $\partial \mathbf{B}/\partial t = 0$, even under transient conditions. Accordingly, curl $\mathbf{E} = 0$ and curl $\mathbf{v} = 0$ are in tandem given by virtue of Eq. (4.4) and the basic coupling law $\mathbf{v} = k\mathbf{E}$. A further crucial simplification immediately drives us to write

$$\mathbf{f}_m = \mu_0 \mathbf{J} \times \mathbf{H} \tag{4.15}$$

because div $\mathbf{v} =$ div $\mathbf{E} = 0$. It is worth observing how the residual magnetic component of $\mathbf{f}_m$ appearing in Eq. (4.15) holds normal to $\mathbf{v}$, so that also the relevant work $\rho_m(\mathbf{f}_m \cdot \mathbf{v})$ [see, for example, Eq. (3.32)] is subject to vanish. In any case, such a transversal component is far too small to exert any significant change of the route guided by the Laplacian pattern. Incidentally, the *a priori*-assumed simplifying condition $\partial \mathbf{B}/\partial t = 0$ practically accounts for the interdiction on $\mathbf{v}$ to acquire a nonzero vorticity, namely, to exert any departure of ion trajectories from the Laplacian pattern. At last, setting $\mathbf{f}_m = 0$, which means methodically zeroing each addend (the force components) of the starting Eq. (4.9), turns out to be an approximating exercise that causes an error within the uncertainties affecting the adopted approximate model. An attractive consideration is that the above trivial result for $\mathbf{f}_m$ is corroborated, even though by a somewhat different reading, in Section 3.11.3 with reference to an incompressible and irrotational flow under rigorously stationary conditions. In the same context, it was verified how the physical abstraction represented by the d'Alembert paradox gains a practical signification when applied to steady and even unsteady real flows once the discussed quasi-static approximation is of course attested.

4.3.6   Unified Momentum and Energy Conservation Law

Bear in mind that $\mathbf{f}_m =$ curl $\mathbf{v} = 0$ and even $(1/\rho_m)$ grad $p =$ grad$(p/\rho_m) \cong 0$ because $p/\rho_m \cong$ const is a tolerable approximation for an equation of state applied to real gases under ordinary physical (ambient and excitation) conditions, often verified in engineering applications (see Appendix 3.A). Therefore, Eq. (3.28) becomes

$$\frac{\partial \mathbf{E}}{\partial t} + k^2 \, \text{grad}\left(\frac{E^2}{2}\right) = 0 \tag{4.16}$$

since $\mathbf{v} = k\mathbf{E}$. Of course, Eq. (4.16) reduces to

$$\operatorname{grad}\left(\varepsilon\frac{E^2}{2}\right) = 0 \qquad (4.17)$$

under steady conditions.

Note that Eq. (4.17) is indistinguishable from $\operatorname{grad}\left(\rho_m\frac{v^2}{2}\right) = 0$ which is the reduced version of $\operatorname{grad} U = \operatorname{grad} T_k = 0$, with $T_k = \rho_m\frac{v^2}{2}$, deduced from Eq. (3.36) in Chapter 3 when $W_T = 0$. Bear in mind that the quantities ε, ρ_m, and k are nonzero constants and the above-mentioned Eq. (3.36) is merely indicative of a zero acceleration condition, formally expressed by $D\,\mathbf{v}/Dt = 0$. Therefore, in a unified manner we envision the electrostatic energy $\varepsilon(E^2/2)$ stored in the gap and the kinetic energy $\rho_m(v^2/2)$ of a steady flow of charges in gases perfectly correspond to each other, at least in a formal sense. It is instructive to dwell for a little while upon the possibility of reproducing the above electrostatic energy distribution even for a fluid at rest with the stipulation that the Laplacian $\mathbf{E}$-field is subject to appropriate boundary conditions. These could be expressed in terms of a discontinued electric field distribution $\mathbf{E} = \mathbf{v}/k$, thus applicable over the injecting surface as well, where $v = kV/L$ (remember that V and L represent the voltage applied to and length of the generic streamline, respectively). Since v is constant along any trajectory, it is immediate to verify that the time of flight, τ, of drifting ions obeys the simple law $\tau = L/v = L^2/(Vk)$ and, therefore, typically holds for the order of 1 m/s (see the closure of Section 3.12.2).

4.3.7 Mobility in the Context of a Coupled Model

With reference to the paired notion of reduced charge mass, the definition of mobility k has been formalized in Section 4.2 through Eq. (4.1). Further considerations on this crucial parameter are also reported in Section 4.3.1. It is well known that the equality $\mathbf{v} = k\mathbf{E}$, with k assumed as a constant proportionality coefficient, is extensively adopted in the realm of electromagnetism and verified by experiment. The practical prerequisite for k to be distinctive of a given ion species is that the low-field assumption is applicable. This is the case when the gas holds in thermal equilibrium, which implies that the electrically imparted translational energy of the charges must be negligible in comparison to the thermal energy that activates chaotic molecular motions and causes random collisions. Under such circumstances, the mean free path λ is irrespective of being the gaseous medium at rest or in motion. Note that this condition appears expressed in Chapter 3 (see, in particular, Section 3.2) in advance to the fluid dynamic treatment of air. A recommended check could be that of using the strong inequality $\lambda \ll K_b T/(Ee)$ which qualifies, after

Wannier (1953), as the low-field state (notation e adopted for the elementary charge). Under the described circumstances, the translational energy gained by the ions along a line segment of length λ is far below the thermal energy of the buffer gas. It is easy to verify that under usual electrostatic and ambient conditions pertaining to this class of problems, λ scarcely approaches 6×10^{-8} m and, hence, is at least one order of magnitude less than the compared ratio.

A redundant but instructive argument can be advanced to corroborate the assumption $k = \text{const}$, especially in consideration of the specialized solenoidal and circulation-free flows under examination. Hence, consider that the ratio q/m and the parameter ζ involved in Eq. (4.1) represent the constant density ρ_0 per unit mass and collision period of a molecule, respectively, in that strictly connected to the mean free path $\lambda = \lambda(1/n)$, where n is the number density of the molecules, the factor ζ is rather sensitive to pressure and temperature changes. However, λ may be estimated to be constant, without incurring an appreciable error, with the distinction that the domain under study is thermally insulated (as previously admitted) and the flow is incompressible and irrotational (see also later on). Under such practically reproducible circumstances, assigning a constant value to k—and, hence, a linear relationship of the kind $\mathbf{v} = k\mathbf{E}$—results in a legitimate settlement (see also Chapter 2). When applied to corona-originated drifting ions, the acknowledged nearly constant character of k is also to be taken as an indirect indicator of the subsonic velocity of an ion flow subject to an $\mathbf{E}$-field. In fact, a fluid moves slowly (velocity typically of the order of 1 m/s) under the action of an electric field when its strength meets the above-mentioned low-field condition. Because a flow enlivened by an electric field is also curl-less, viscosity-related frictional mechanisms and, in turn, pressure changes, combined with heat formation and transfer internal to the charged moving gas, become quite unimportant mechanisms. Briefly speaking, the above-claimed constancy of k, intimately connected to the specialized status of incompressible and irrotational flow, is nothing but a distinctive property of ion drifts. Of course, ion mobility is a property of excited matter; therefore, any relationship between k and viscosity μ is legitimately expected. Conceptually, k and μ are in inverse ratio so that setting $k\mu = \text{const}$ turns out to be an educated guess. Indeed, finding an explicit form for the k–μ relationship is not a strictly required task in the present treatment, even because k and μ are in practice singly constant, throughout. However, it cannot be passed over that this correlation is burdened with a certain symbolic meaning in the context of a coupling model, perhaps because the parameters involved are traditionally pertaining to and distinctive of decomposed sections of physics. Consider that the notion of ion mobility, so extensively adopted in electrodynamics, electroaerodynamics, physical chemistry, atmospheric physic, and molecular biology to list a few, is, to the authors' knowledge, rather extraneous to the contents of classic fluid-dynamic textbooks. On the other hand, the bibliography devoted to electro- and/or magnetohydrodynamics seems defective in understanding the explanatory character of, and hence clearly showing, any direct connection between k and μ (unless the fluid is a liquid, in which case the approximate Walden's rule $k\mu = $ const is adopted under low-field conditions [see Walden (1906)]). In order to fill this void, the roles that such quantities interplay in a joined model are estimated in Appendix 4.C.

The inherent inconsistency of the equality $\mathbf{v} = k\mathbf{E}$ with a Poissonian structure of the $\mathbf{v}$- and $\mathbf{E}$-fields and, conversely, its inherent consistency with Laplacian fields has been carefully discussed in Section 4.3.4. The validity of the above formula can also be proved differently if reference is made to the potentials φ_V and φ_E simultaneously assumed by a charged particle moving with velocity $\mathbf{v}$ along a streamline. This specific subject is, for the sake of convenience, carefully dealt with in Appendix 4.D.

4.3.8   Some Remarks on the Deutsch Hypothesis (DH)

According to this simplifying assumption, subject to ongoing criticism with a long history, the fluxline patterns of $\mathbf{E}$ and the corresponding space-charge-free field $\mathbf{E}_L$ hold indistinguishable (Chapter 2). In other words, the ion-flow trajectories (streamlines) are admitted as being rigidly guided by the Laplacian fluxlines. Therefore, the notions of streamline and fluxline, as well as those of streamtube and fluxtube, are geometrically indistinguishable from each other, even though the prefixes "stream" and "flux" apply here to moving-mass-related and immaterial fields, respectively. As a formal consequence, $\mathbf{E} = \ell\mathbf{E}_L$, with the enhancing dimensionless factor ℓ being a scalar function of position. The previous investigation has carefully ascertained that the sole reproducible $\mathbf{E}$-field picture compatible with a subsonic ion drift agrees, as a matter of fact, with DH, even though it shows lack of transversal continuity. However, it seems instructive to insist on the validity of DH by introducing further theoretical arguments directly involving the geometry of the ion paths. If $\delta\mathbf{l}$ denotes an oriented segment associated with a linear particle moving with velocity $\mathbf{v} = k\,\mathbf{E}$, then spatiotemporal changes of such an elementary quantity are governed by the equation

$$\frac{D(\mathbf{l})}{Dt} = (\delta\mathbf{l} \cdot \text{grad})\mathbf{v} = (\text{grad } \mathbf{v}) \cdot \delta\mathbf{l} \tag{4.18}$$

The above material particle is to be conceived as having an elementary length $\delta\mathbf{l}$ and negligible, thus molecular at least, cross-sectional dimensions. Any theoretical discussion on whether the orientations of $\delta\mathbf{l}$ and $\mathbf{E}$ can mutually differ or not should be made according to Eq. (4.18) [see Landau, Lifschitz, and Pitaevskii (2008)]. On the other hand,

$$\frac{D\mathbf{v}}{Dt} = \frac{\partial v}{\partial t} + (\text{grad } \mathbf{v}) \cdot \mathbf{v} = -\frac{[\text{grad } p + \mu \text{ curl}(\text{curl } \mathbf{v})]}{\rho_m} + \mathbf{f}_m \qquad (4.19)$$

derives from rearranging Eqs. (3.14) and Eq. (3.26). The specialized flow examined here allows Eq. (4.19) to be zeroed (see the beginning of Section 4.3.6). Namely, we can claim with measured pedantry that in a thermally insulated ambient crossed by an irrotational and solenoidal ion flow the isothermal compressibility α and bulk expansion β coefficients (see Appendix 3.A) substantially represent constant attributes of the fluid material along any given streamline. An immediate implication is that pressure p and temperature T behave likewise. Accordingly, it is proved that both $\partial \mathbf{v}/\partial t$ and grad $\mathbf{v}$ drop to zero [zero acceleration, see Eq. (4.19) above and the previously discussed Eq. (4.16)] and Eqs. (4.18) and (4.19) in turn assume the specialized forms

$$\frac{D(\delta \mathbf{l})}{Dt} = 0 \qquad (4.20)$$

and

$$\frac{D\mathbf{v}}{Dt} = 0 \qquad (4.21)$$

Equations (4.20) and (4.21) show a similar structure, which importantly implies that the rates of change of $\delta \mathbf{l}$ and $\mathbf{v}$ are identically zero as the electric field $\mathbf{E} = \mathbf{v}/k$ is obliged to hold tangential to $\delta \mathbf{l}$ throughout the drift region. The latter performance is also related to the applied boundary conditions, previously discussed in Section 4.2.4, according to which the ion injecting velocity $\mathbf{v}$ is tangential to the Laplacian fluxlines, in turn orthogonally emerging from the source. The substantial result is that each **E**-line in the local presence of moving space charge may be identified with a flowline. Using an attractive terminology after Alfvén, ion swarm and **E**-field appear as being "frozen together," thus showing a certain resemblance with some magnetohydrodynamic performances applied to perfectly conducting plasmas [see, for example, Moffatt (1978)]. Distinctive of ion-drift electrohydrodynamics is that a linear particle of length $\delta \mathbf{l}$ is unaffected by both spreading (transversal expansion) and longitudinal stretching/shortening during the motion along preestablished Laplacian routes. Such constraints fully prevent ion flow from being a space-filling one, which is configured as a nonuniformly dense aggregation of filamentary channels individually guided by as many Laplacian electric-field lines, as will be substantiated by experimental data furnished under especially clarifying laboratory circumstances (Chapter 5),

- The space-charge-free electric field due to charge residing in electrostatic equilibrium on the surface of the electrode system is uninfluential and inessential for assessing the "frozen-in" **E**-field. In fact, the latter is generated by current-carrying channels under steady conditions, so that the interspace field filled by particles at rest is also affected.
- Determination of the trajectory pattern is the sole preserved portion of the stably shaped actual source, exactly as if the corona-free parts of the same active conductor were substantially deprived of electrostatic influence.

Indeed, no violation of DH occurs in using the raised model since the **E**-field either way happens to be frozen into ion-swarm filaments guided by an $\mathbf{E}_L$-field pattern, even though this can differ from the one under completely corona-free conditions. Differently speaking, the electrostatic situation is that the ionized cloud not only shields the inner features of the conductor adopted as an injector, but also largely prevails as a field source over the inactive surface areas of the same living conductor. The combined substantial effect is that the true Laplacian pattern of the drifting trajectories is determined by the shapes of both the steady charge cloud and the *a priori* established opposite collector.

Returning now to the above discussion strictly involving Eqs. (4.20) and (4.21), it not only supplies subsidiary rigorous reasons in defense of DH, but also represents a favorable argument for the self-consistency of an ion swarm's filamentary structure. In fact, the given direct ratio between $\delta \mathbf{l}$ and $\mathbf{v}$ (or $\mathbf{E}$) theoretically proves that any elemental current-carrying channel seemingly behaves as a filamentary conductor along which the electrodynamic field is unidimensional and uniform. In Section 4.5.2, the above arguments are sustained following a logical *reductio ad absurdum* route, while the subsection just below is devoted to the field calculation according to the discussed filamentary performances.

To better appreciate the limits for DH applicability, let supersonic dynamics apply in the extended field domain for mere explanatory reasons. Consider that this is nothing more than a realistic admission for the class of problems under investigation. Accordingly, the condition div $\mathbf{v} =$ div $\mathbf{E} = 0$ goes off and the discontinued structure of the **v**-field can no longer be sustained. As a result, the thin channels settle down in a compact structure, thus macroscopically forming smoothed-out streamtubes with varying cross-sections and bending with no regard for the Laplacian pattern. The established **E**-field will be continuous and Poissonian and, therefore, will be theoretically inconsistent with DH. Consider further that DH fails even while admitting that the additional force component $\mu_0\,\mathbf{J}\times\mathbf{H}$ impressed to the supersonic flow [see Eq. (4.14)] persists in being faint enough for appreciable changes of the ionic routes to occur. Under the described circumstances and according to Eq. (4.14), the nonzero body force becomes $\mathbf{f}_m=\rho\mathbf{E}$ since ε_0 div $\mathbf{E}=\rho$.

4.4   DISCONTINUOUS ION-FLOW PARAMETERS

4.4.1   Multichanneled Structure

The discussed uniformity conserved by ρ and v along each trajectory is manifestly incompatible, as somewhere and by some means introduced in advance, with a continuous field structure. Under the latter circumstances, the walls of adjacent streamtubes remain joined throughout, even when they are forced to longitudinally expand/contract in an appropriate manner. Therefore, in order to preserve the ρ and v conservation, each elemental streamtube of an ideally smoothed-out field representation must split into an unpredictably large number of conducting channels of constant and macroscopically negligible cross-sectional area da. The mutually disjoined filamentary channels (physical streamlines) forming a widespread, discrete ion swarm crossing the drift region (Section 4.3.4) are superimposed to the corresponding fluxlines of a given Laplacian pattern. Responsive for this performance is that the per-unit kinetic energy

$$e_k=\frac{\rho_m v^2}{2} \tag{4.22}$$

examined in Chapter 3, Section 3.12 by a variational approach is required to be as small as it can be, thus exactly in conformity to DH. With reference to the ith channel of length L_i, in which charged particles of mass density $\rho_{m,i}$ and charge density ρ_i drift with velocity v_i, the associated kinetic energy becomes

$$E_{k,i}=\frac{1}{2}\rho_{m,i}v_i^2 L_i da=\frac{p_k\rho_i}{L_i} \tag{4.23}$$

Note, in particular, that $\rho_{m,i}$ and, in turn, ρ_i depend on L_i, while the constant p_k, which is equal to $k^2V^2 da/(2\rho_c)$, is proportional to the single cross-sectional area da. Extending the minimization to $E_{k,i}$, namely, zeroing the partial derivative of Eq. (4.23) performed with respect to L_i, ultimately gives $\rho_i/L_i=$const. The ratio V/L_i represents the magnitude of the electric field E_i internal to the ith channel carrying the elemental current $\mathscr{I}_i$. Accordingly, it is a simple exercise to realize how the current density J_i and, in turn, $\mathscr{I}_i$ are invariants of the multichannel model ($i = 1 \div N$). In fact, $J_i = k\,\rho_i E_i = kV\,\rho_i/L_i$ and $\mathscr{I}_i = J_i da = \mathscr{I}$, so that the pair of constant quantities J_i and $\mathscr{I}$ unambiguously apply to the single N channels. In addition,

$$I = N\mathscr{I} = kVN\frac{da\rho_i}{L_i} \tag{4.24}$$

is the formula for the total corona current. Because of the constancy of J_i and $\mathscr{I}$ even for a general field presenting a curvilinear pattern, it seems permissible to invoke the existence of a current-equivalent field model whose streamlines are straight and, therefore, of the same length. In particular, use will be made in Section 4.4.2 of the concentric-electrode geometry with the final intent of gaining somehow an insight into the boundary conditions for the unknown current distribution in the corresponding real domain. The voltage drop in the plasma region is comparatively unimportant, even because this region usually occupies a negligible fractional amount of the entire gap. All things considered, adopting an injection surface that perfectly fits the active conductor results in a permissible assumption leading to set $E_i = V/L_i$. This is a good approximation for the uniform electrodynamic field located throughout the unidimensional ith channel, having been respectively designed with V and L_i the voltage applied to the electrode system and length of the channel-guiding Laplacian fluxline.

4.4.2 Current Distribution

Consistent with the discussed multichanneled ion-swarm model is the distribution law

$$\left(\frac{J}{J_0}\right)_e = \left(\frac{dA_0}{dA}\right)_c \tag{4.25}$$

for the average current density $(J)_e$ at the emitter. In this regard see Section 4.3.3 (and related Appendix 4.A) where more detailed considerations, claiming usage of average quantities, are made. When referring to field quantities, the adopted symbol $(\!-\!)_{e,c}$ designates the average value over the cross-sectional area $(dA)_{e,c}$ of any given elemental streamtube, whose inner channels are guided by the Laplacian fluxlines. To each elementary streamtube/ fluxtube corresponds an assigned cross-sectional area $(dA)_e$ attached to the outer ionization region of the active conductor. The tubes impact the opposite electrode with differently dilated areas $(dA)_c$, in the sense that the ratio $(dA)_c/(dA)_e$ generally changes from a streamtube to another, depending on how the generically curvilinear field is structured. If an elementary streamtube convoys a number of channels equal to dN, then the channel number density $(n)_e$ at the emitter is such that $(ndA)_e = dN$ and, therefore, $(J/J_0)_e = (n/n_0)_e$ since $(J)_e = (n)_e \mathscr{I}$ and $(J_0)_e = (n_0)_e \mathscr{I}$. The above-mentioned simplified condition interpreting the ionization edge as being super-imposed to, thus substantially identified with, the electrode surface derives from a pair of combinative conditions: The thickness of this plasma region is negligible in comparison to the depth of the ion-drift region; the diffusion layer peripheral to this region causes, unless differently stated (see Section 4.3.1), unimportant deformation of the ionization edge with respect to the inner shape of the active electrode. The former condition is always respected, while the latter is a very specialized one because it occurs when diffusion is negligible, namely, when the ionization occupies a steady layer evenly surrounding the surface of the active electrode. Notably, the wire-plane setup represents a paradigmatic case in which the outer surface of the corona sheet is a coaxial cylinder.

The quantities subscripted with 0, which are distinctive of the elementary streamtube having shortest length L_0, are adopted as referential ones for the normalization of the involved distribution laws. The final formula for the referential current density at the emitter reads

$$(J_0)e = \frac{kV(V - V_T)(dA_0)_c}{(dA_0)_e L_0^3} \qquad (4.26)$$

where V_T designates, according to a practical interpretation of Kaptzov's law, the corona voltage theoretically just over, thus substantially undistinguishable from, the onset level. Equation (4.26) descends from the discernible relation-ship $dI_0 = dN_0 \mathscr{I} = (J_0 dA_0)_e = (J_0 dA_0)_c$, where dI_0 is the ion current carried by the referential fluxtube and

$$(J_0)_c = \frac{I}{2\pi\varepsilon L_0^2} \qquad (4.27)$$

with

$$I = \frac{kV(V - VT)2\pi\varepsilon}{L_0} \qquad (4.28)$$

The present model gives substantial arguments to vindicate the general character of both Eqs. (4.27) and (4.28), the latter being interpreted as an applicative reformulation of Eq. (4.24). In fact, the discovered property $\rho_i/L_i = \text{const}$ evokes an unsophisticated axial-symmetric representation (field-pattern transformation; see Figure 4.1), from which Eq. (4.28) can be readily obtained. The inner radius r is left unspecified, while the outer

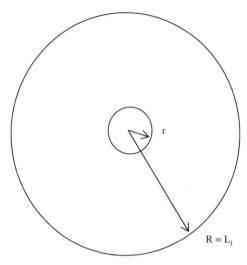

FIGURE 4.1 Axial-symmetric transformation for calculating the general corona current.

radius L_i is provisionally set equal to the length of the ith trajectory chosen at random in the true ion-flow pattern.

Using some simple manipulations, the equality

$$N\rho_i da = 2\pi\varepsilon(V - V_T)\xi_a, \quad \text{where} \quad \xi_a < 1 \tag{4.29}$$

is ultimately given. By hindsight, consider that ξ_a is a dimensionless factor less than but approaching unity, especially when the trajectories of the original flow are closer together. Therefore, ξ_a depends on the value assigned to L_i, in the sense that the longer the value of L_i, the smaller the value of ξ_a. Setting $\xi_a = 1$ turns out to be a permissible approximation, provided that the outer radius $L_i = L_0$, namely, L_i, is as small as it can be. In other words, ξ_a gets close to unity wherever the interchannel spacing is reduced even if the channels tend to mutually depart. Under such conditions, the total field in the neighborhood of a given streamline of length L may be represented by a smoothed-out one approximately equal to $E = V/L$, throughout. The circumstances described in practice apply to filamentary flows of ions with the specification that L is limited, insofar as the flow is preserved from exceeding dilution due to overly divergent Laplacian fluxlines. To appreciate the value of Eq. (4.29), consider that the terms on both sides of the equality represent different formulations of the same per-unit charge released in the spacing between concentric surfaces of radii r and $R = L_i$. This space extracharge appears when the applied voltage V exceeds the onset level V_T. In fact, there is a certain amount of charge, proportional to V_T and irrespective of V, that resides in electrostatic equilibrium on the active electrode. With specific reference to the quantity on the left-hand side of the above equation, consider that the total space charge $N\rho_i L_i da$, physically concentrated in an assembly of N outwardly directed L_i-long channels, has been divided by the length, also equal to L_i, of the elemental cylindrical model. The corresponding space charge per unit length on the right-hand side is to be interpreted as being distributed in the gap. In this matter, $\varepsilon(V - V_T)\xi_a/L_i$ is the charge density on the outer surface, the difference $V - V_T$ being in accord with Kaptzov's law. As previously said, the multiplicative factor ξ_a, less than but tending to unity, takes into account the not exactly radial structure of the interchannel electric field. Tacitly assuming the equalities $\rho_i/L_i = \rho_0/L_0$ and $\xi_a = 1$ for the reasons also discussed above, then rearranging Eqs. (4.24) and (4.29), gives Eq. (4.28). In the light of the above treatment, Eq. (4.27) appears immediately discernible. It is worth noting how Eq. (4.28) fully reminds us of the celebrated Townsend formula for the corona current in general cases. Consider that previous investigations attained a similar result with minor success since, for example, it was assumed that V_T is negligible in comparison to V. Even though permissible in several practical cases, the need for making recourse to the above restrictive assumption is rather indicative of more important deficiencies in the theoretical substrate. As often occurs, the error attenuates as far as the calculation is restricted to integral quantities. This is the case for the corona current, notably for pioneer works on coronas which were preferentially dedicated to this integral parameter than to its distribution in the drift region (perhaps, even in the reasonable expectation of insuperable, or at least long-lasting, difficulties that unrestricted investigations would encounter in the future). The general character of Eq. (4.28) leads to postulate that once the actual ambient conditions are established, V_T unambiguously assumes a single, minimum value associated with the most protruding location.

This is the right time to urge a theory for the diffusion layer positioned just inside the ionization cloud. In light of the previous preparatory mentions, Eq. (4.25) represents the definitive boundary condition for the current density of just-generated ions prior to their transversal injection into the drift region. However, Eq. (4.25) strictly depends on the stable configuration assumed by the plasma-filled ionization region masking, in both the electrostatic and optical senses, the material conductor's inner features. In this respect, the outermost diffusion layer can assume a key role in shaping the actual emitting surface and, therefore, in affecting the Laplacian pattern on which Eq. (4.25) ultimately depends through the ratio $(dA_0/dA)_c$. For the same reason, not only J, but all the field parameters implied this investigation (see also Appendix 4.A), are influenced by the disfigured stable shape that the injecting surface can assume. This surface, interfaced between the diffusion layer and the outer drift region, covers in whole or in part that of the high-voltage conductor. The described field distortion significantly extends up to the collector, although the plasma region and collector are often widely apart. Also consider that the laws governing, in particular, the distributions of J, E, and ρ at the emitter are of prominent interest because some of them need to be involved in the form of boundary conditions for computational purposes. For this exposition to proceed in an orderly fashion, the entire treatment leading to a final formulation for the above-mentioned trio of laws at the emitter, and not only there, moves to Appendix 4.A. As advised in Section 4.3.1, the practical value of them is of course subordinated to the knowledge of the definitive shape assumed by the peripheral diffusion layer. This specification refers to the fact that the original problem of assigning appropriate boundary conditions in terms of electrical quantities becomes that of wisely establishing the actual morphology of the source.

4.4.3 More on the Average Quantities

For the reader's convenience, the equations for the calculation of average electric-field parameters—current density, charge density, and electric field—are summarized below. Using physical quantities averaged over the small cross-sectional area dA of a fluxtube/streamtube is twice as important for the model: Direct comparisons with both previous theories and available experimental data are made permissible. In fact, the supplied average quantities evoke the notion of a smoothed-out field, a representation largely adopted elsewhere; in addition, detected data are generally given averaging over probe surfaces that are large in comparison to the molecular cross-sectional area da of each filament composing the intercepted multifilament flow (see Chapter 5).

Known variables are the onset voltage V_T as well as the geometrical quantities represented by the pairs $(dA)_e$, $(dA)_c$ of the elementary surfaces at the opposite ends of an elemental streamline of length L. Of course, even the referential quantities $(dA_0)_e$, $(dA_0)_c$, and L_0 are established. It is worth considering that the above geometrical quantities are distinctive of a Laplacian pattern and, therefore, can be supplied by a customary exercise because of the large availability of computational resources applied to Laplacian fields, often equipped with subsidiary graphical facilities. Even V_T can be calculated using the same standard code, thus precisely finding a given applied voltage $V = V_T$ capable of determining somewhere a surface electric field equal to the threshold E_T. This minimum ignition value, changing in function of ambient conditions and going around the standard value of $24\,kV/cm$, is of course localized on the overstressed point(s) of the emitter surface [see Hartmann (1984)]. Therefore, the set of useful formulas at the emitter are as follows:

- $(J/J_0)_e = (dA_0/dA)_c$ with $(J_0)_e = kV(V - V_T)[(dA_0)_c/(dA_0)_e L_0^3]$ for the calculation of the average current density [see Eqs. (4.25) and (4.26)].

- $E_e = E = V/L$ for the calculation of the average electric field, with L being denoted the length of the median streamline (see below in this subsection for additional details).
- $\rho_e = J_e L/(k \cdot V)$ for the calculation of the average charge density because $J_e = k\rho_e E$ with $E = V/L$ (see Section 4.3.1).

The Laplacian pattern of the ion swarm allows average values to be calculated for any generic point of the field region, namely, for each section dA of an elementary streamtube whose median length is L. The weighty task of carefully calculating the electric field in the interchannel spacing only filled by neutrals can be spared. According to a smoothed-out envisioned model of the drifting ion flow (see also Appendix 4.A), it is reasonable to expect that the interchannel electric field surrounding an individual intrachannel of length L is slightly departed from the value V/L. This pleasant performance especially manifests in the regions where the average current density intensifies, hence at or even at some distance from the emitter. In fact, the channels invariably tend to gather, and therefore to restore the transversal uniformity of the electric field at the expenses of the longitudinal one, exactly in such critical regions. As related to the distribution $(dA_0/dA)_c$ and lengths L (see the relevant bulleted statements above), the calculation of $(J/J_0)_e$, E, and ρ_e, however, demands preliminary assessment of the Laplacian-field pattern subject to a full identification of the source. Regarding this concern, consider that the additional field due to the charge deposited in electrostatic equilibrium on the electrodes is assumed to be insignificant in a comparative sense. This remark neglects the presence (in general, it is the case) of the inactive parts of the injecting conductor and, hence, to envision the confined source as being in isolation. On the other hand, though, the source shape is *a priori* unknown in general because it can remarkably depart from that of the inner material electrode. As will be shown in due course, this fairly complicated original problem is, however, solvable by implementing an approximate method based on shape smoothing (see also Section 5.9).

4.5 DEPARTURES FROM PREVIOUS THEORIES

It has previously been appreciated how the arguments put forward by this procedure offer a coupled theoretical framework for an appropriate treatment of the unipolar ion-drift problem. The key discovery for the given model to be claimed is that the electric current of a true ion drift, by definition continuously obstructed by a neutral mass, can be undistinguishable from that of an untrue convection in an ambient devoid of buffer mass. Accordingly, the former current is interpretable as due to an equivalent migration of reduced-charge-bearing molecules of reduced mass to which a subsonic velocity $\mathbf{v}$, numerically equal to $k\mathbf{E}$, is impressed. The condition of incompressibility div $\mathbf{v} = $ div $\mathbf{E} = 0$ appears safely applicable to the physical ion drift, provided that above-mentioned equivalence is invoked.

The next subsection will be aimed at theoretically showing the most striking differences between previous issues (see Chapter 2) and the present coupled approach. In the same context, some easily discernible inconsistencies affecting the former theoretical predictions will be analyzed. This gives additional substantiation to the unified model where, instead, the above-mentioned difficulties disappear altogether. A description and interpretation of related experimental data, supplied by methodical laboratory activity on special electrode assemblies, will form a separate subject to which Chapter 5 is entirely dedicated.

4.5.1 Ion-Drift Formulation

A decoupled electromagnetic approach leads to the celebrated ion-drift equation

$$\frac{D\rho}{Dt} = -\frac{k}{\varepsilon_0}\rho^2 \tag{4.30}$$

expressed as a rate of change of ρ. Equation (4.30) can also be derived from Eq. (4.2) after considering the identity div$(\rho\mathbf{v}) = \rho$ div $\mathbf{v} + \mathbf{v}$ grad ρ by div $\mathbf{v}$ replaced by $k\rho/\varepsilon_0$. It is immediately verified that this ratio is obtained by arranging Gauss's law and the equality $\mathbf{v} = k\mathbf{E}$. Before going on with this description, it should be remarked incidentally that, once the electrode geometry is assigned, $\mathbf{E}$ can be envisaged as the superposition of a purely Laplacian field determined by the applied voltage $\mathbf{V}$ and a field that establishes when the space charge of local density ρ is held motionless and the electrodes are short-circuited. Resolving Eq. (4.30) for ρ gives

$$\rho(t) = \rho_0\left[1 + \frac{k\rho_0}{\varepsilon_0}t\right]^{-1} \tag{4.31}$$

with $\rho_0 = \rho(t = 0)$. Granted, for the sake of argument, that the drifting trajectory is known, the line distribution of ρ can be obtained starting from Eq. (4.31), otherwise, Eq. (4.31) is substantially unfruitful. Under the so-called saturated or, differently defined, space-charge-dominated conditions formally expressed as $1 \ll (k\rho(t = 0)/\varepsilon_0)t$, Eq. (4.31) reduces to the ρ_0-independent asymptotic formula $\rho(t) = (\varepsilon_0/kt)$. Therefore, a saturated ρ-distribution is connected to the concept that $\rho(t=0)$ or t tends to infinity, in which case the electric field contributed by the space charge largely prevails over the space-charge-free component. In practice, the above theoretical strong inequality is satisfied when $V \gg V_T$, in which case the electric field **E**, substantially contributed by the sole space-charge component, is also expected to slightly change along any given trajectory. In fact, massive quantities of space charge are invariably prone to arrange in such a way as to attenuate field gradients. Under such circumstances, the time-dependent Eq. (4.31) could turn into the space-dependent formula

$$\rho(\ell) = \frac{\varepsilon_0 E}{\ell} = \frac{\varepsilon_0 v}{k\ell} = \frac{\varepsilon_0 V}{\ell L} \tag{4.32}$$

so as to give

$$J(\ell) = k\rho(\ell)E = k\frac{\varepsilon_0 V^2}{\ell L^2} \tag{4.33}$$

Here, $\ell = vt$ represents the position coordinate along the path taken by an ion drifting with approximately constant velocity $v = kE = kV/L$. In reality, the investigators themselves are restrained from safely adopting Eqs. (4.32) and (4.33) unless $\ell = L$, in which case ρ and J only refer to the collecting electrode. Such a recommended practice is understandable—for reasons carefully discussed in Section 4.3.2 and tested by an example of application—but quite arbitrary, thus being indicative of the basic deficiencies affecting uncoupled models, irrespective of their degree of sophistication. The practical validity attributable to the above pair of equations depends on the fact that the trajectory of length L is *a priori* assigned or easily calculable because of special facilities assumed for the **E**-field pattern. This occurs when use is made of Deutsch's hypothesis, according to which L is the length of a Laplacian fluxline. Parenthetically, the ions' time of flight τ is given by resolving the above formula $\rho(\tau) = \varepsilon_0/k\tau$ for τ, where $\rho(\tau) = \rho(L) = \varepsilon_0 V/L^2$. At last,

$$\tau = \frac{L^2}{Vk} \tag{4.34}$$

is easily derived.

On the other hand, our coupled model causes div **v** to vanish, so that the previously adopted identity $\mathrm{div}(\rho\mathbf{v}) = \rho\ \mathrm{div}\ \mathbf{v} + \mathbf{v}\ \mathrm{grad}\ \rho$ reduces to div $(\rho\mathbf{v}) = \mathbf{v}\ \mathrm{grad}\ \rho$ and, in turn, Eq. (4.2) becomes $\partial\rho/\partial t + \mathbf{v}\ \mathrm{grad}\ \rho = D\rho/Dt = 0$. In other words, the charge density is expected to hold constant in space (along any given trajectory, but changing in general from a trajectory to another) and in time (unsteady conditions prevented). This result is quite different from the predictions given through Eqs. (4.31) and (4.32), even imposing $\ell = L$ in the latter formula. In addition, Section 4.3.8 advises that the often-cited Deutsch's hypothesis really deserves to be reevaluated so that it is allowed to gain the rank of governing law. In this way, $\rho = JL/(Vk)$ with J subject to the boundary conditions expressed by Eqs. (4.25) and (4.26). Owing to the same reason, only Eq. (4.34) is shared, with the secondary distinction that this formula is rigorous in the coupled model and approximate in the decoupled one.

In Chapter 3 (see Section 3.12), the energy density of the bulk motion was identified as a purely kinetic energy $T_k = \frac{1}{2}\rho_m v^2$ owing to the vanishing additional energy density $W_t = W + p$. This total positional term is contributed by the potential energy due to external and internal forces which, as a matter of fact, tend to zero when reference is made to an incompressible and irrotational ion flow. Under such specialized circumstances, ρ_m is constant, as is the modulus of the velocity $\mathbf{v}$, along a fluxline. In hindsight, consider that T_k could at first appear with all but the electrostatic energy density $\frac{1}{2}\varepsilon_0 E^2$ up to the multiplicative constant $\rho_m k^2/\varepsilon_0$ (such a correspondence was previously mentioned in Section 4.3.6). The excuse for this surprising observation will be fully cleared in the next subsection.

4.5.2 Comparative Discussion

Several treatments adopt DH as a simplified means to resolve generally decoupled ion-drift problems. On the other hand, it was carefully verified in Section 4.3.8 that DH is invalid only owing to its inherent incompatibility with Poissonian fields. Overlooking this inconvenience implies a tacit admission, according to Gauss's law, that div $\mathbf{v} = k$ div $\mathbf{E} = k\rho/\varepsilon_0$ differs from zero. That being stated, use is again made hereafter of the key relationships previously adopted in Section 4.3.8, with the very intent of showing how exploring the existence of limits for the DH application, when div $\mathbf{E}$ differs from zero, is an unfruitful exercise. Accordingly, Eq. (4.18) and Eq. (3.26) can, respectively, be rewritten

$$\frac{D(\delta\mathbf{l})}{Dt} = (k/\varepsilon_0)\rho\delta\mathbf{l}, \quad \frac{D\mathbf{E}}{Dt} = (1/k)\rho\mathbf{E} \tag{4.35}$$

Even though similarly structured, Eqs. (4.35) cannot ensure now that the vector $\delta\mathbf{l}$ and $\mathbf{E}$ applied on a common field point are coincident, namely, that field fluxlines and ion trajectories are, according to DH, superimposed. However, provided that $\delta\mathbf{l} = \upsilon\mathbf{E}$ with υ a proportionality factor distinctive of a given trajectory (or fluxline in this case), then Eqs. (4.35) become reciprocally indistinguishable to within an inessential constant. Therefore, $\mathbf{E}$ and $\delta\mathbf{l}$ are equally oriented and their moduli are in direct ratio along any trajectory. But this is the case only if $\mathbf{E}$ and $d\ell$ are individually constant along any fluxline, or, differently speaking, if the field is an electrodynamic and solenoidal one. As an immediate consequence, the right-hand sides of both Eqs. (4.35) need to be *a priori* zeroed in order for Eqs. (4.20) and (4.21) to be restored. Of course, the difficulties raised clear up replacing the questioned continuous field with the discontinuous, multifilamentary Laplacian field recommended in this chapter. In fact, the formal presence of a nonzero ρ in Eqs. (4.35) is a conceptual absurdity, with such a presence being inconsistent with the prerequisite for a solenoidal and irrotational field to exist, or, differently speaking, for DH to hold.

It is worth noting that putting $\delta\mathbf{l} = \upsilon\mathbf{E}$ seems, at first sight, to be compatible with specialized Poissonian $\mathbf{E}$-fields whose fluxlines are straight. This is because the magnitude E is prejudicially believed to hold constant, compared to constant linear increments $d\ell$, especially along a straight trajectory. Unfortunately, even this subsidiary expectancy turns out to be illusory, even under space-charge-dominated conditions, since the constant value of E claimed is exactly incompatible with a Poissonian field. As a convincing argument, consider the analytically amenable case study treated hereafter.

As an illustrative example, let us pay attention to the space-charge electric field $\mathbf{E}$ ahead of the one-ended indefinitely long conducting filament depicted in Figure 4.2. This physical abstraction is in the reality approached when a straight thin wire carries a steady current i_c. This current is intended as the prerequisite for a unipolar corona activity to be sustained at the terminal denoted by $r = 0$ (from now on, notation ℓ is replaced with r usually evoking a radius). A saturated space charge is expected to fill the entire drift region

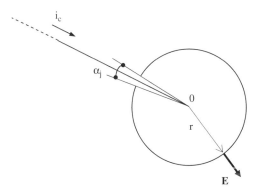

FIGURE 4.2 E-field surrounding the end of a current carrying semi-infinite filament. The spherical equipotential of generic radius r nearly approaches the filament (cone angle α_j deliberately exaggerated; applied voltage neglected). This theoretical scheme indifferently applies to space-charge-free and space-charge-filled domain case studies. Because of the impressed current i_c, the field sources respectively are the charge accumulating at, or spherically spreading up from, the origin 0.

because the condition $V \gg V_T$ is easily met on the terminal. The current spreads up at the wire terminal, thus according to the same spherical symmetry of the corresponding Laplacian field. Note that the latter is the theoretical result of charge accumulation at the origin $r = 0$, an occurrence determined by i_c and V_T when they are respectively impressed and infinitely large. For the sake of precision, such a field abruptly vanishes in a very restricted conical region, coaxial to and surrounding the wire, whose acute cone apex is positioned on the wire terminal (cone angle $\alpha_j \to 0$). Therefore, legitimately taking advantage of neglecting the above detail, as well as existence of a restricted ionization region, simply gives

$$dQ = 4\pi r^2 \rho(r)\, dr = 4\pi \varepsilon_0 E \rho dr \qquad (4.36)$$

According to the first equality in Eq. (4.32), Eq. (4.36) represents the charge of density $\rho = \varepsilon_0 E / r$ which happens to fill the elementary spherical shell of radius r and thickness dr. Consider that

$$i_c = \frac{dQ}{dt} = kE\frac{dQ}{dr} = 4k\pi\varepsilon_0 E^2 r \qquad (4.37)$$

since $dr/dt = v = kE$, with E expressing, as previously repeatedly specified, a nearly constant space-charge-dominated field magnitude. Of course, this approximation for E is permissible in the surroundings of the origin, namely, until r denotes an unspecified finite quantity. Notwithstanding this reasonable restriction, Eq. (4.37) tacitly manifests, through the presence of the multiplicative variable r, an irreconcilable violation of the current continuity. In fact, i_c is predicted to increase while passing through progressively dilating spherical cross sections of radius r. Note that this absurd result is especially disappointing in consideration of the presumptively successful straight-fluxline case study just now examined.

In a commendable analysis conducted by Sigmond (1986) and extensively reported in Chapter 2, all DH-based methods have been first categorized according to the termed longitudinal L- and transversal T-type formulas and then criticized. However, the adopted models without distinction accommodate DH, usually expressed in the form of $\mathbf{E} = \ell \mathbf{E}_L$, in a set of physical equations also taking into account Gauss's law div $\mathbf{E} = \rho/\varepsilon_0$, current continuity div $\mathbf{J} = 0$, and field irrotationality curl $\mathbf{E} = 0$. The cited investigator proves by eloquent arguments that the field calculation invariably suffers from inconsistencies and shortcomings when use is made of such models. Briefly, L-type models conserve current flow guided by Laplacian fluxlines but violate orthogonality with the equipotential surfaces in the gap; T-type models conserve the orthogonal Laplacian pattern but violate current conservation [by the way, the above example of application leading to Eq. (4.37) is categorized as a T-type model]. Hence, he concludes the following:

1. The raised violations are evocative of the unphysical character of DH, even though preference should be given to L-type models preserving current continuity because they are prone to supply better current distributions, as often required for practical applications.
2. Even using sophisticated numerical algorithms, the solutions are compromised by ambiguities in the phase of problem specification. In other words, the current distribution at and geometry of the interface between the ionization region and drift region are still arbitrary. These form the boundary conditions to which the above methods are subject.

Contrary to observation 1, the present unified model shows that the so-called DH is all but a different way to state that setting div $\mathbf{E}$ = rot $\mathbf{E}$ = 0 is distinctive of practical electro-fluid dynamic circumstances. Since nothing prevents such combined field features from being physically admissible, then DH instead deserves to be identified as a physical law. Importantly, note further how the arguments raised in Section 4.3.4, as well as the boundary condition given in Section 4.4.2 for the current density [see Eqs. (4.25) and (4.26)], wipe out the difficulties raised in observation 2.

Last, even the energy density T_k described at the end of Section 4.5.1 deserves to be singled out for special consideration. The joined model gives $W_E = \rho = 0$, so that $U = T_k = \frac{1}{2}\rho_m v^2 = (\rho_m k^2/\varepsilon_0)\frac{1}{2}\varepsilon_0 E^2$, thus in contradiction with an uncoupled phenomenological envisioned event where U instead represents the per-unit energy stored in the $\mathbf{E}$-field [see, for example, Jones (1992)]. As $U = \frac{1}{2}\varepsilon_0 E^2 = T_k + W_E$, the intimate and exclusive connection, claimed by the unified model, between kinetic and electrostatic energies breaks up. In fact, as will be appreciated later on, the quantity T_k is left unspecified in uncoupled models, whereas the added term $W_E = \frac{1}{2}\rho\varphi_E$ is identified as the electrostatic potential energy density. Note that W_E is equal to half the per-unit total work $\rho\varphi_E$, the potential φ_E being of course such that $\mathbf{E} = -\text{grad}\ \varphi_E$, for the charging process to be accomplished. The missing per-unit work $\rho\varphi_E - W_E$ is, therefore, still equal to $\frac{1}{2}\rho\varphi_E$ but represents a supplementary energy cost to simultaneously polarize the molecule according to a composite electronic-orientational polarization process. Any attempt to interpret the nature of T_k is circumvented in favor of a variational approach aimed at finding an explicit form for Lagrange's function. This task is easily executed by performing the simple manipulation $L = T_k - W_E = U - 2\ W_E$ so that $L = \frac{1}{2}\varepsilon_0 E^2 - \rho\varphi_E$ is ultimately given. In conformity with this procedure, $W_E dV_m$ and $q = \rho dV_m$ respectively represent the work of charging an individual molecule of volume dV_m and related charge q imparted by the charging process. Once again, a theoretical difficulty arises because of the same basic weakness: a questionable broken up model is adopted in substitution of a unified self-consistent counterpart. Regarding this, let the existence of a nonzero quantity $W_E dV_m$ be *ab absurdo* admitted to charge the molecule of mass m and surface area dA_m. Under such circumstances,

$$qE = \frac{m}{\zeta} v \qquad (4.38)$$

where

$$q = c_m \varepsilon_0 E dA_m \qquad (4.39)$$

In Eq. (4.38), v/ζ represents the rate of change of the average velocity of a charge-bearing molecule of mass m acted upon by the electrostatic force qE. This interpretation descends from the key presence of the collision period ζ previously introduced in Eq. (4.1). On the other hand, the structure of Eq. (4.39) basically conforms to those adopted in describing a quick charging process in order for the molecule to acquire the saturation charge q (parenthetically, care should be taken with regard to the notion of saturation applied here to a single molecule; see also Section 4.2, which differs from that adopted in Section 4.5.1 with reference to space-charge-dominated conditions). Note the presence of the unspecified dimensionless coefficient c_m taking into account the relative permittivity of, and Knudsen's number associated with, the molecule. Introducing Eq. (4.39) into Eq. (4.38) immediately gives

$$v = \frac{c_m \varepsilon_0 \zeta \, dA_m}{m} E^2 \qquad (4.40)$$

The error affecting the above approximate procedure is not detrimental for a first-order interpretation of Eq. (4.40) according to which the velocity depends on the electric field squared through the constant proportionality ratio $c_m \varepsilon_0 \zeta \, dA_m / m$. This formula clashes with the linear relationship $\mathbf{v} = k\mathbf{E}$ corroborated by theoretical and experimental data and, therefore, safely applicable to subsonic (low-field conditions) ion flows (see Sections 4.2 and 4.3.7). Especially with reference to the subject of Section 4.3.7, the pair of kinematic conditions represented by zero vorticity and zero expansion appears tacitly synthesized by the simple formula $\mathbf{v} = k\mathbf{E}$. This is to stress that using the above equality with no restriction for div $\mathbf{E}$, as invariably found elsewhere, results in a serious theoretical incongruence. According to this questioned formulation, any charging process aimed at restoring the energy density $\frac{1}{2}\varepsilon_0 E^2$ by the work of charging W_E appears applicable and, hence, the notion of charging work undertakes a theoretical meaning. Additionally, note how Eq. (4.39) is governed by Gauss's law, thus contrary to the coupling approach. Instead, in our model, Gauss's law is interdicted for the reasons carefully discussed in advance, and the quantity equal to $\frac{1}{2}\varepsilon_0 E^2$ is rather evocative of a kinetic energy acquired by the ions *externally* to the drift region than an electrostatic energy stored *internally* to the same domain. It has been said several times that the above reasoning applies to the large drift region in which the ion velocity is subsonic. In contrast, in the active glow region, things change dramatically as has often been said here, in the sense that the velocity is expected to become supersonic. This implies that a variational approach rather in keeping with, say, with Jones (1992) and Jones (2000) can be granted.

4.5.3 Ionic Wind in the Drift Zone

The very physical mechanism giving rise to the corona-originated electric wind still remains a challenging research topic. Basically, use is made of the Navier–Stokes equation

$$\rho_m \frac{Dv_m}{Dt} = -\mathrm{grad}\, p + (\mu + \mu')\mathrm{grad}\,\mathrm{div}\,\mathbf{v}_m + \mu\nabla^2\mathbf{v}_m + \rho_m\mathbf{f}_m \qquad (4.41)$$

conveniently rewritten here [see Eq. (3.25)] with some quantities fitted to the situation. Specifically, ρ_m and $\mathbf{v}_m$ respectively now express the gas's density and velocity; the presumed nonzero charge density ρ is incorporated in the Coulomb force $\mathbf{f}_m = \rho\mathbf{E}$. This electric body force derives from rearranging Eq. (4.14), deprived of the unimportant magnetic component, and Gauss's law $\mathrm{div}(\varepsilon\mathbf{E}) = \rho$. Commercially available software is often adopted to evaluate the $\mathbf{v}_m$-field once $\mathbf{f}_m$ is predetermined by a separate calculation of the **E**-field in the presence of space charge of density ρ. It is clear that, according to this model, the ionic wind, represented by $\mathbf{v}_m$, is generated inside the drift region where Coulombian forces are presumed to impart air motion by a momentum-transfer mechanism. The plasma is often neglected as a possible companion source of airflow, presumably because of the zero-net (or approximately so) charge density in the corona zone.

Returning to Eq. (3.25), the present treatment clearly invites us to inquire otherwise into the formative cause of the impinging air jet. Concisely, we can say that in the drift zone the following statements are true:

- $\mathbf{f}_m$ vanishes (see Section 4.2.5).
- Equation (4.30) ultimately reduces to $D\mathbf{v}/Dt = 0$ by virtue of the discussed identification of $\mathbf{v}_m$ with $\mathbf{v}$ (see Section 4.2), subject to the laws $\mathbf{v} = k\mathbf{E}$ and $\mathrm{div}\,\mathbf{v} = 0$.

Therefore, conversely to previous predictions [see, for example, Adamiak and Atten (2004)] and in consideration of both (a) some discordances with experiment [this is the case when application is made to lifters, as raised by Chen, Rong-de, and Bang-jao (2013)] and (b) some limitations of the adopted computational codes [rotating machinery applications; see Lemire, Vo, and Benner (2009)] raised elsewhere, it is recommended here to pay careful attention to the concentrated ionization zone and its periphery, rather than the drift zone, as an effective site of sudden momentum transfer by ion-to-neutral collisions. The ion velocity in a plasma is largely supersonic, which implies that the key quantities div $\mathbf{v}$ and $\mathbf{f}_m$ both differ from zero. Additionally, even though the charge density averaged throughout the ionization region vanishes, the special structure of an elemental avalanche and the complex charge morphology of the overall plasma invites us to estimate the total force $\mathbf{f}_m$ as a nonzero one. Therefore, the drifting zone is inferred as being only crossed by, while the ionizing zone actually blows, the ionic wind. The air "pumping" mechanism in the ionizing zone still needs to be fully cleared (see later); however, in support of this zone as an airflow source, consider that important ion wind performances are being investigated exactly in close proximity to active conductors, thus inside the ionization zone and at the interface with the outer drift region where the wind is presumably generated. This is clearly evidenced, for instance, in avionic and turbomachinery studies finalized to prototyping plasma actuators. Once such corona discharge-based air-moving devices are mounted, some important aerodynamic effects, still needing to be carefully examined do take place throughout and on the periphery of the plasma region where the airflow is produced.

Air molecules trapped in the plasma are presumed to be repeatedly pushed into the drift region owing to the pulsative nature of several corona-discharge modes. The voltage-dependent repetition rate of the discharge is large enough to allow the elemental jetting pulses to merge in time, namely, the overall air pumping to appear steady at distanced detection points. In fact, the ionic wind is usually measured far away from the ionization region, thus in the drift region, so that the plasma activity is not influenced. In this passive region, highly directional airflows happen to be concentrated exactly where the streamlines gather according to the above theory; at the same time, these lines could be intended as natural continuations, in the extended drift region, of the original tiny avalanches composing the active region. Therefore, the more or less prominent directional character detected for such a neutral airflow is transitively argued to depend on the local compactness degree of the individual avalanches forming the plasma discharge. These elemental filaments are in general distributed according to a specific law ultimately dictated by the geometry of the electrode assembly. When the pair of surfaces bounding the plasma are allowed to assume a concentric configuration, then the plasma will look evenly compact.

However, a detailed description of the air jet genesis in the plasma still needs to be definitively furnished. An intuitive momentum transfer mechanism could imply the heads of elongating avalanches as outgoing pushers for the neutral mass. An alternative, more sophisticated mechanism could involve inter-avalanche compressive/depressive forces according to the elsewhere-cited cardinal law of electrodynamics (Pappas, 1993). With reference to this composite law, the total force $\mathbf{F}_{12}$ acting, say, on the charge q_1 moving with velocity $\mathbf{v}_1$ in the presence of a charge q_2 moving with velocity $\mathbf{v}_2$ at distance $\mathbf{r}_{12}$ from the former obeys the relationship

$$\mathbf{F}_{12} = \frac{q_1 q_2}{4\pi\varepsilon_0} \frac{\mathbf{r}_{12}}{r_{12}^3} \left[1 + \frac{3}{(r_{12}c)^2} (\mathbf{v}_1\mathbf{r}_{12})(\mathbf{v}_2\mathbf{r}_{12}) - \frac{2}{c^2} \mathbf{v}_1 \cdot \mathbf{v}_2 \right] \qquad (4.42)$$

Unity in brackets is related to the Coulombian component, while the remaining pair of terms are related to the simultaneous Amperian component. For explanatory reasons, Figure 4.3 regards the case in which q_1 and q_2 are twin charges (as to polarity and value) moving with the same velocity $\mathbf{v}$ along parallel trajectories placed at distance r_{12}, the latter also being the distance between the charges. Accordingly, Eq. (4.42) assumes the specialized form

$$\mathbf{F}_{12} = \frac{q_1 q_2}{4\pi\varepsilon_0} \frac{\mathbf{r}_{12}}{r_{12}^3} \left[1 - 2\left(\frac{v}{c}\right)^2 \right] \mathbf{u} \qquad (4.43)$$

where $\mathbf{u}$ is a unit vector also reported in the figure. Note that $\mathbf{F}_{12}$ becomes *positive*, thus repulsive, if $v < c/\sqrt{2}$ or becomes *negative*, thus attractive, if $v > c/\sqrt{2}$. In the first case the Coulombian force prevails against the Amperian counterpart, whereas in the second case the situation reverses.

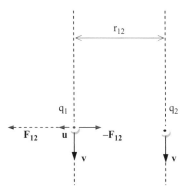

FIGURE 4.3 Twin charges q_1 and q_2 moving side by side with velocity $\mathbf{v}$. The force $\mathbf{F}_{12}$ specifically acting on the charge q_1 is oriented as, or in opposition to, the unit vector $\mathbf{u}$ if the Coulombian component prevails the Amperian one or vice versa, respectively.

Equation (4.43) is indicative of the fact that charges moving along contiguous lines can be effective in alternatively causing mutual attraction and repulsion among the avalanches. This overall mechanism is determined by the pulsating nature of the original plasma current and the acknowledged ability of the implied moving charge to even attain a velocity equal to $c/\sqrt{2}$ and over. The relevant mechanical phenomenon could therefore result in a total pushing effect, longitudinally directed on the neutral mass as a result of transversal alternating compression, which fills the interpulse spacing. Of course, the above-mentioned longitudinal ion wind pumping is expected in any case because of the diverging, rather than parallel, directions often assumed by single contiguous avalanches. In fact, active regions are exactly localized in correspondence of overstressed large-curvature regions of conducting surfaces from/toward which fluxlines radially diverge/converge.

4.6 CONCLUDING REMARKS ON THE LAPLACIAN STRUCTURE OF ION FLOWS

The following deserves to be highlighted (see Appendix 4.A):

- Equation (4.A.1) [or (4.A.2), indifferently] and, in turn, Eq. (4.25) exactly apply when the flow field is discontinuous, namely, composed of a discrete number of Laplacian streamlines.
- The pattern serving, let us say, as a template for a filamentary ion swarm to be reconstructed in shape and density, is made up of Laplacian fieldlines.
- The Laplacian pattern allows the boundary conditions at the emitting surface to be unambiguously and fully expressed in the form of normalized distribution laws for J, E, and ρ. Therefore, reaping profit from the treatment skipped in Appendix 4.A,

$$\left(\frac{J}{J_0}\right)_e = \left(\frac{dA_0}{dA}\right)_c; \left(\frac{E}{E_0}\right)_e = \frac{L_0}{L}; \left(\frac{\rho}{\rho_0}\right)_e = \left(\frac{dA_0}{dA}\right)_c \frac{L}{L_0} \qquad (4.44)$$

collectively derives. Incidentally, note that Eq. (4.A.11) tells us that the current conservation along any elemental fluxtube acts, in agreement with the key distribution law dN/dN_0, in determining the homeomorphism that $(J/J_0)_e$ shares with $(E_L/E_{L_0})_e$ and, as expected to a lesser extent, with $(\rho/\rho_0)_e$.

- Taking a look at Eqs. (4.A.12)–(4.A.14) enables some prompt forecasts to be made, especially after admitting $(E_{L_0}/E_L)_e = 1$. In short, $(J/J_0)_c$ is expected to change faster than $(\rho/\rho_0)_e$ because the space charge really tends to uniformly fill the interelectrode gap as much as it can. Inherently linked to this performance is that the geometry of the electrode assembly might have restrained impact on E/E_0, thus differently to what happens to $(E_L/E_{L_0})_c$, especially in the absence of corona activity at the counterelectrode. On the other hand, the space-charge motion is significantly affected by the above-mentioned geometry [and $(E_L/E_{L_0})_c$ itself through Eq. (4.25)] insofar as the current distribution purely depends on the ratio dN/dN_0 through Eq. (4.A.11). On this issue, and in view of Eq. (4.A.2), it has been verified that such a distribution law operates better as a sensitive surrogate of the Laplacian pattern than the competing ratio L_0/L (see Appendix 5.A for additional remarks, provided in the light of general invariance principles, regarding the key ratio dN/dN_0).

- In principle, the redundant set of Eqs. (4.A.12)–(4.A.14) can be exploited to resolve the well-posed inverse source problem of reconstructing ion-injection properties at the emitter, meaningfully represented by the ratio $(J/J_0)_e$, by randomly using one of the distributions $(E/E_0)_c$, $(J/J_0)_c$, or $(\rho/\rho_0)_c$ measurable at the distanced collector. The problem's well-posedness basically depends on the stressed homeomorphism, so that the unpleasant and impairing task of performing direct detection of source features is circumvented. Being especially practicable and reliable, remote measurement at the passive electrode is strongly recommended for gaining insight into the corona performances. Indeed, special reasons connected to the overall validity of the given measurements will lead, by elimination, to the preferred use of Eq. (4.A.12) alone (see Section 5.1 in Chapter 5 for detailed considerations on the subject).

- The above-mentioned experimental approach can prove compellingly necessary in the present investigation. This is because the ion injecting surface is often prone to assume an asymptotic shape, generally different from the geometry of the material emitter. This complicates matters in the lack of an available supportive theory capable of *a priori* providing the actual Laplacian pattern represented by the $(E_L)_e$, $(E_L)_c$, and L distributions.

- In the case of 2D fields, the generic fluxtube's cross section dA collapses into a line element, thus lying on a Laplacian equipotential curve. Of course, such a line element is oriented along the direction which is the one and only for N to change at a maximum rate. Nothing of importance is lost by restriction of the generic 3D treatment to that applicable to 2D symmetric fields, with the advantage to make it analytically amenable and, therefore, invaluable for a physical interpretation of the results. It is with regard to these very issues that the experimental investigation reported in Chapter 5 is ultimately addressed to some informative 2D case studies. As beneficial consequence, an instructional cue device to preliminarily establish the shape of the actual source is also provided in that chapter at the end of Section 5.9. The practice now being discussed has some claims to circumvent a supplementary rigorous but involved estimation of the diffusion layer geometry. The advice reported here-under in Appendix 4.B does enable the reader to clearly forecast the raised difficulty.

APPENDIX 4.A

Arguments in favor of the crucial formula

$$\left(\frac{n}{n_0}\right)_e = \left(\frac{dA_0}{dA}\right)_c \qquad (4.A.1)$$

have deliberately been left over in Section 4.4 in fairness to the reader. Note that when combining Eq. (4.A.1) with the equality $(n/n_0)_e = (J/J_0)_e$ already discussed in Section 4.4.2, then Eq. (4.25) is obtained. Again with reference to Eq. (4.A.1), what often happens is that the ratio dA_0/dA at the collector cannot be influenced dramatically by the shape of the material electrode on which a corona source is somewhere positioned. This in part could depend on the often found considerable distance of the collector from the emitting points but first and foremost on the shielding performances of the ionization region, in the sense that in general it rather behaves as a prevailing and deformed field source, even in the presence of inactive parts of the same electrode. The pointed or wedge-shaped active locations are submerged by an overall ionized cloud, thus behaving as an actual source showing off an even out asymptotic morphology substantially meeting the condition $(E_L/E_{L,0})_e = 1$ (Appendix 4. B). Ultimately, even $(J/J_0)_e$ is somehow affected, through Eq. (4.A.1), by the described performance since $(dA_0/dA)_c$ is directly correlated to the Laplacian correspondent of $(E_L/E_{L,0})_e$ at the collector [see Eq. (4.A.3)]. An alternative formulation of Eq. (4.A.1) is

$$\left(\frac{dA_0}{dA}\right)_e \frac{dN}{dN_0} = \left(\frac{dA_0}{dA}\right)_c \qquad (4.A.2)$$

because $(dA_0/dA)_e \cdot dN/dN_0 = (n/n_0)_e$. Bear in mind that N, dN, and dN_0 are finite numbers representing, respectively, the totality of drift channels composing the multichanneled ion flow and those convoyed into two elementary fluxtubes that are orthogonally attached to the corresponding surfaces $(dA)_e$ and $(dA_0)_e$ of the emitter. For the sake of generality, $(dA)_e \neq (dA_0)_e$ where subscript 0 pertains to a referential fluxtube. Invoking E_L-flux and current continuities along the above-mentioned fluxtubes respectively gives

$$\left(\frac{dA_0}{dA}\right)_e \left(\frac{dA}{dA_0}\right)_c = \left(\frac{E_L}{E_{L,0}}\right)_e \left(\frac{E_{L,0}}{E_L}\right)_c \qquad (4.A.3)$$

and

$$\left(\frac{dA_0}{dA}\right)_e \left(\frac{dA}{dA_0}\right)_c = \left(\frac{J}{J_0}\right)_e \left(\frac{J_0}{J}\right)_c \qquad (4.A.4)$$

In the light of Eq. (4.A.2), the above pair of formulas can be rewritten as follows:

$$\left(\frac{E_L}{E_{L_0}}\right)_e \frac{dN}{dN_0} = \left(\frac{E_L}{E_{L_0}}\right)_c \qquad (4.A.5)$$

$$\left(\frac{J}{J_0}\right)_e \frac{dN}{dN_0} = \left(\frac{J}{J_0}\right)_c \qquad (4.A.6)$$

Therefore, Eqs. (4.A.5) and (4.A.6) turn out to be examples of a unique homeomorphism (continuous transformation) because the correspondence between the double pair of dimensionless quantities $(E_L/E_{L_0})_e$, $(E_L/E_{L_0})_c$ and $(J/J_0)_e$, $(J/J_0)_c$ is one-to-one (i.e., a bijection), continuous in both directions, and expressed by the multiplicative factor dN/dN_0 as a transforming function. It is worth noting that the transformation that is being discussed vouches for the soundness of the coaxial model reported in Section 4.4.2 and represented in Figure 4.1. To better appreciate this statement, and passing over the formal pedantry required, consider that there is nothing to prevent Eq. (4.A.2) from being alternatively written as

$$\left(\frac{dA dN}{dA_0 dN_0}\right)_c = \left(\frac{dA}{dA_0}\right)_e \qquad (4.A.7)$$

This is because $(dN/dN_0)_c = (dN/dN_0)_e = dN/dN_0$ is distinctive of any given elementary fluxtube convoying dN channels compared to a referential fluxtube with dN_0 inner channels. An immediate reading of Eq. (4.A.7) is one that the space arrangement of the totality of N channels allows the collector surface to be partitioned into a number of elements. The surface area of an individual element is expressed by $(dA \cdot dN)_c$ to within an unspecified trivial constant, and the relevant normalized distribution law $(dA \cdot dN)_c/(dA_0 \cdot dN_0)_c$ at the collector resembles the corresponding one $(dA/dA_0)_e$ at the emitter. Similar considerations could be repeated for the vector quantities $\mathbf{E_L}$ and $\mathbf{J}$ with the stipulation, consistent with Eqs. (4.A.5) and (4.A.6), that both transformations of the physical distributions E_L/E_{L_0} and J/J_0 at the collector are multiplied by the ratio dN_0/dN instead of its inverse. Having said that, an axial-symmetric representation of the original field under examination appears feasible to the extent that the notion of transformed elementary surface $(dA \cdot dN)_c$ is virtually introduced as a substitute for the physical one $(dA)_c$. How the homeomorphism applies to the average charge density ρ remains yet to be seen. In this matter, note that a structure of the formula quite similar to that given for Eqs. (4.A.5) and (4.A.6) can again be applied to ρ, namely,

$$\left(\frac{\rho}{\rho_0}\right)_e \frac{dN}{dN_0} = \left(\frac{\rho}{\rho_0}\right)_c \tag{4.A.8}$$

provided that a certain degree of approximation is admitted. In fact, it is a simple exercise to verify that Eq. (4.A.8) derived from Eq. (4.A.6) after involving the relationship $J = k\rho E$ and in view of the equality

$$\left(\frac{E}{E_0}\right)_e = \left(\frac{E}{E_0}\right)_c \tag{4.A.9}$$

For the sake of precision, the approximate value of Eq. (4.A.8) must be pointed out as soon as the conservation of E along individual Laplacian fluxlines is extended to elementary Laplacian fluxtubes, which is the case here. On the whole, what matters to ultimately highlight is that the raised connection between the above homeomorphism and the coaxial model described further back in Section 4.4.2 proves the soundness of Eqs. (4.A.1) and, in turn, (4.25). Incidentally, note that Eq. (4.A.9) differs from the joint structure of Eqs. (4.A.5), (4.A.6), and (4.A.8) because the quantity approximately conserved along an individual Laplacian fluxtube is $E = V/L$ instead of its flux $E \cdot dA$. In fact, rearranging Eqs. (4.A.5) and (4.A.9) gives

$$\frac{(E_0 \cdot dA_0)_e}{(E_0 \cdot dA_0)_c} \frac{dN}{dN_0} = \frac{(E \cdot dA)_e}{(E \cdot dA)_c} \tag{4.A.10}$$

In general, dN/dN_0 differs from unity and, in consequence, the flux $E \cdot dA$ changes along any given elementary fluxtube of median arclength L. For the sake of comparison, note that with reference to the *JdA* flux, which identifies with the current *dI*,

$$\left(\frac{dI}{dI_0}\right)_c = \left(\frac{dI}{dI_0}\right)_e = \frac{dI}{dI_0} = \frac{dN}{dN_0} \qquad (4.A.11)$$

can be written in view of Eqs. (4.25) and (4.A.2).

Equations. (4.A.6), (4.A.8) and (4.A.9) easily lead us to write

$$\left(\frac{J}{J_0}\right)_c = \left(\frac{J}{J_0}\right)_e \left(\frac{E_L}{E_{L_0}}\right)_c \left(\frac{E_{L_0}}{E_L}\right)_e \qquad (4.A.12)$$

$$\left(\frac{\rho}{\rho_0}\right)_c = \left(\frac{J}{J_0}\right)_e \frac{L}{L_0} \left(\frac{E_L}{E_{L_0}}\right)_c \left(\frac{E_{L_0}}{E_L}\right)_e \qquad (4.A.13)$$

$$\left(\frac{E}{E_0}\right)_c = \frac{E}{E_0} = \frac{L_0}{L} \qquad (4.A.14)$$

usefully expressed as a function of the boundary condition $(J/J_0)_e = (dA_0/dA)_c$ (for reasons which will fully be appreciated in Chapter 5) and Laplacian-field-related dimensionless quantities.

APPENDIX 4.B

Neglecting ion diffusion, other than local ionization and chemical reactions, authorizes the conservation law

$$\frac{\partial n}{\partial t} + \mathrm{div}\,(n\mathbf{v}) = 0 \qquad (4.B.1)$$

for the number density n to be looked upon [see also Eq. (3.A.1)]. Owing to the combinative notion of reduced mass m and charge q introduced in Section 4.2, the above formula also holds after methodically multiplying the involved members by constant factors, such as q or m. Therefore, Eqs. (4.2) and (4.3) respectively are derived after performing such specialized multiplications.

Introducing now the diffusional contribution in Eq. (4.B.1) gives

$$\frac{\partial n}{\partial t} + \mathrm{div}[n(\mathbf{v} + \mathbf{v}_D)] = 0 \qquad (4.B.2)$$

in which $\mathbf{v}_D$ denotes diffusion velocity impressed by the driving force grad n to the just generated space charge surrounding the ionization region of the active electrode. Accordingly [see, for example, Beuthe and Chang (1995)],

$$\mathbf{v}_D = -D\frac{\text{grad } n}{n} \tag{4.B.3}$$

clearly shows how a nonzero grad n is, through the diffusion coefficient D, the prerequisite for a local concentration of generated charges to spread up with velocity $\mathbf{v}_D$. For a gas in thermal equilibrium, the ratio D/k approaches, according to Einstein's formula, the quantity KbT/e (e stands for elementary charge). After performing the divergence of the sum in brackets and rearranging with Eq. (4.B.3), Eq. (4.B.2) can be rewritten as

$$\frac{\partial n}{\partial t} + \mathbf{v}\,\text{grad } n - D\nabla^2 n = 0 \tag{4.B.4}$$

Note that the additional condition div $\mathbf{v} = 0$ for subsonic flows and the identity $\text{div}(\text{grad } n) = \nabla^2 n$ have been tacitly adopted. According to Eq. (4.B.2), $\mathbf{v}$ is in principle expected to largely prevail over $\mathbf{v}_D$ (and its components) wherever $\mathbf{v}$ differs from zero in the drift domain. However, grad n drops practically to zero along any individual filamentary streamline, so that the field is forced to remain unidimensional and Laplacian, being unaffected by charge accumulation (otherwise causing a nonzero grad n). In other words, for an elongated, incompressible particle traveling down with subsonic velocity $\mathbf{v} = k\mathbf{E}$, a streamline guided by a Laplacian pattern causes, in particular, practical interdiction to diffusion along such a route. Not only this, but also the interfilament spacing, where $\mathbf{v} = 0$, can be considered as being nearly preserved from diffusional invasion of the recently generated space charge (see below).

Returning now to the Laplacian structure of the **E**-field, the following observations require supplementary attention before discussing Eq. (4.B.4) and a related physical interpretation:

- As previously discussed, charge transfer throughout the filamentary channels composing the ion swarm is largely determined by a drifting, rather than diffusional, mechanism (see also the end of this appendix, where an estimation of Péclet's number, associated with the filamentary flow, is provided).
- Owing to existence of a presumably very large number of drift channels composing the ion swarm, even the **E**-field performances in the outer spacing among channels cannot significantly differ from those assumed along them. Of course, such a prediction, extended to the spacing outside

the multichannel ion swarm, is especially sustainable in close proximity to the emission zone where the channels converge. In that crucial, restricted zone surrounding the active emitter, the space-charge-free **E**-field outside the compact assembly of drift channels can hardly depart, in magnitude and orientation, from the one "frozen" in the channels (see Section 4.3.8).

This being considered, the Laplacian and one-dimensional nature somehow conserved by the **E**-field everywhere just beyond the ionization layer causes the outwardly directed grad n to vanish. This implies that Eq. (4.B.4) reduces to the well-known Fick's (second) law

$$\frac{\partial n}{\partial t} - D\nabla^2 n = 0 \qquad (4.B.5)$$

with the specification that Eq. (4.B.5) only applies tangentially to the emission surface. In other words, the diffusion of a space-charge cloud of volumetric density n, governed by Eq. (4.B.5), can be thought of in terms of superficial expansion that "licks up" the emitter surface. This behavior occurs in the absence of the component traversing the drift region along a substantially uniform electric field. Accordingly, only a material transport identified as ion drift in the **v**-field, the velocity **v** being equal to $k\mathbf{E}$, is permissible transversally to the injecting surface. Note that the charge diffusion is obliged, after a transient period of time, to remain "squashed" to the above surface, even beyond the confined zones where ionization take place. In other words, larger and larger space portions in the surroundings of the above-mentioned active zones become impenetrable to diffusion transversally to the conductor surface. This happens because new drifting channels, transversal to the expanding diffusion boundary layer, are continuously formed. As the voltage increases, the overstressed surface zones where the original ionization sources are positioned tend to hold the originally confined extension owing to the shielding action of charges previously diffused all around. It is worth bearing in mind that charge injection by corona activity in general exhibits a pulsating character. Therefore, the apparently steady nature observed for the glow corona and, irrespective of corona mode, for the drift current detected at the collector is instead the result of a pulse merging over timescales pertaining to the ion's time of flight τ and time of diffusion τ_D. We stress that the diffusion layer and the transversal charge injection pattern are prone to become stable on the periphery of the ionization region and in the drift region, respectively. Therefore, the structure of Eq. (4.B.5) needs, in general, specification of appropriate initial and boundary conditions. Provided that the charge injection exerted by each individual pulse is theoretically represented by the Dirac delta function, then the solutions of Eq. (4.B.5) are expressed by integrals of exponentials (exponential or error functions, etc.). In this regard, some further considerations, only based on first-order estimations of the diffusion and drift timescales, can be helpful. The time τ_D over a certain distance L_D for a superficial diffusion is equal to

$$\tau_D = \frac{L_D^2}{pD} \tag{4.B.6}$$

and rather approaches the order of 10^{-2} s. For this estimation, L_D and D have been reasonably assumed to be of the order of 10^{-3} m and 10^{-5} m²/s, respectively, while p has been set equal to 4 ($p=2, 4, 6$ for, respectively, 1D, 2D, and 3D diffusions). Therefore, τ_D is often one order of magnitude less than $\tau = L/v$ (time of flight τ of seconds along any representative streamline of length L), which confirms that Eq. (4.B.5) often applies to a diffusion layer as being immune from transversal subtraction of ions to be convoyed into drift channels. As a result, there is time enough for the driven ions to assume a stable distribution throughout the mean-square distance L_D calculated on the surface of the emitter. This distribution tends to mask and smooth out the detailed features (asperities, edges, etc.) of those parts of the material electrode which are submerged by the ionized cloud. Differently speaking, the injecting surface of the actual ion source is prone to assume an asymptotic geometry radiused away from the active sections of the inner electrode. Only under uncommon circumstances, in which a particular geometry of the active electrode is combined with an exceptionally high voltage applied, could τ_D and τ become comparable quantities. The consequence is that the ion injection from the actual source can be conversely affected by a prominent concentration. In any case, with reference to a common distance represented by a streamline length L [i.e., L_D is now set equal to L in Eq. (4.B.6)], alternatively use can be made of Péclet's number $P_e = \tau_D/\tau = Lv/(2D)$. For the sake of precision, p becomes equal to 2 owing to the unidimensional nature of the transfer considered in the present calculation. Because $P_e \gg 1$ indicates drift more efficient than diffusion (the contrary for $P_e \ll 1$), one can affirm with the help of additional arguments that diffusion is quite unimportant along a streamline, with P_e being of the order of $10^4 \gg 1$ in this case.

As a closure of the above subject regarding diffusion-originated stable shaping of outer injection surfaces, it is helpful to remind ourselves that a plasma layer behaves as a conducting sheet that could mask, in an electrostatic sense, the features of the inner active parts of the emitter. In other words, the dimensionless quantity (n/n_0) is allowed to remarkably differ from that related to the sheer geometry of the living conductor. Of course, computationally responsible for this difference is the pair of distribution laws $(E_L/E_{L,0})_e$ and $(E_L/E_{L,0})_c$ implied in Eq. (4.A.3), both influenced by field distortion because of the actual morphology assumed in each individual case study by the steady injecting surface.

APPENDIX 4.C

According to an ideal model of monomolecular gas in thermal equilibrium,

$$\mu = \frac{1}{3} nm\lambda w \tag{4.C.1}$$

is a possible formulation for the dynamic viscosity μ. This law tacitly refers to spherical molecules having mass m, number density n, mean free path λ, and temperature-dependent average speed w. Equation (4.C.1) can be involved in classical treatments of conduction and hydrodynamic models of charge flows [see, for example, Solymar and Walsh (1993)]. In particular, $w = \lambda/\zeta$ with $\zeta = mk/q$ [see Eq. (4.1) rearranged], so that Eq. (4.C.1) can be rewritten as

$$\mu = nq\frac{\lambda^2}{3k} \tag{4.C.2}$$

As $\lambda^2 = \left[2\left(\pi d^2 n\right)^2\right]^{-1}$ with d expressing the molecular diameter, it is easy to ultimately verify that for a molecule of surface area dA_m the product

$$k\mu = \frac{q}{6ndA_m^2} \tag{4.C.3}$$

equals a constant, provided that n is constant as well. This result has been predicted in Section 4.3.7 exactly with reference to incompressible flows. However, it is instructive to learn that n is, in general, the only variable involved in Eq. (4.C.3) since a charge saturation up to a value q is physically expected for each individual molecule of mass m and an exposed surface of area dA_m. For the sake of comparison, an explicit form of Walden's rule $k\mu = q/6\pi d/2 = \text{const}$ is given, invoking Stokes' law for a charge-bearing carrier in a liquid medium.

APPENDIX 4.D

In the **v**-field we have

$$v = -\frac{\dfrac{\partial \varphi_v}{\partial t}}{\dfrac{\partial \varphi_v}{\partial n_v}} = \frac{\dfrac{\partial \varphi_v}{\partial t}}{\mathbf{n}_V \,\mathrm{grad}\, \varphi_v} = -\frac{\dfrac{\partial \varphi_v}{\partial t}}{\left|\mathrm{grad}\, \varphi_v\right|} \tag{4.D.1}$$

since the unit vector $\mathbf{n}_v$ grad $\varphi_v / |\text{grad } \varphi_v|$ is normal to the equipotential surface $\varphi_V(\mathbf{r}, t) = \text{const}$ and oriented toward φ_V drops. Rearranging Eq. (4.D.1) and bearing in mind that $v = \mathbf{n}_v \cdot \mathbf{n}$ gives

$$\frac{\partial \varphi_v}{\partial t} + \mathbf{n}_V \cdot \mathbf{v} |\text{grad } \varphi_v| = \frac{\partial \varphi_v}{\partial t} + \mathbf{v} \text{ grad } \varphi_v = \frac{D\varphi_v}{Dt} = 0 \qquad (4.D.2)$$

Equation (4.D.2) tells us, as expected, that the local rate of structural change for the function $\varphi_V(\mathbf{r}, t)$, when the fluid particle passes through the point $\mathbf{r}$ at the instant t of time with velocity $\mathbf{v} = d\mathbf{r}/dt$, is zero. It is a straightforward exercise to verify that fully performing the divergence of Eq. (4.D.2) ultimately leads to div $\mathbf{v} = 0$. According to this result, v is constant along any given streamline and changes, in general, in the function of the chosen streamline. Once again, the $\mathbf{v}$-field is verified to identify with a set of unidimensional fields whose individual streamlines participate to set up a discontinuous, filamentary flow of charged fluid. The above proof can likewise be repeated for the electric field $\mathbf{E} = \mathbf{v}/k$, with the relevant equipotential surfaces $\varphi_E(\mathbf{r}, t) = \text{const} = \varphi_V/k$ being superimposed to those of the $\mathbf{v}$-field. Parenthetically, the same procedure can be extended somehow to time-independent space-charge-free electrostatic fields so that Eq. (4.D.2) becomes

$$\frac{D\varphi_E}{Dt} = \mathbf{v}_{0b} \text{ grad } \varphi_E = 0 \qquad (4.D.3)$$

The excuse for the validity of Eq. (4.D.3) is that there is nothing to prevent us from introducing $\mathbf{v}_{ob}$ as the arbitrary velocity of a moving immaterial fieldpoint along any given Laplacian fluxline. A suitable reading of Eq. (4.D.3) is that such a point explorer collectively perceives the infinite succession of equipotential surfaces, all orthogonal to its motion, as the family $\varphi_E(\mathbf{r}) = \text{const}$. The numerical value of the constant is of course demanded to pick out an arbitrary surface. Because no correlation holds between $\mathbf{v}_{ob}$ and $\mathbf{E} = -\text{grad } \varphi_E$, the electrostatic field is free to assume a general three-dimensional structure throughout the domain under examination.

CHAPTER 5

EXPERIMENTAL INVESTIGATION ON UNIPOLAR ION FLOWS

5.1 INTRODUCTION

Among other things, the final bulleted list in Section 4.6 succinctly points out that Eq. (4.25) is distinctive of a discontinuous ion flow, precisely guided by countable Laplacian streamlines, rather prone to be evidenced through a well-thought-out set of laboratory tests reproducing 2D fields. Therefore, this chapter is aimed at being responsive to such an expected substantiation as well as it can. A 3D generalization of the experimentally discovered performances will be attempted in light of the deep connection between the acknowledged general character of the homeomorphism adopted in Section 4.4 (see also Appendix 4.A) and the notion of invariance given in Section 5.9 of the present chapter. Electrode setups are largely employed for gaining an insight into the physical properties of corona activity and associated ion flows. Even though prototypes, full-scale facilities, and small-scale models could of course exhibit complex geometries motivated by practical reasons, basic research does routinely follow a restricted number of unsophisticated paradigms. These can be represented by the wire-plane, point-plane (sometimes termed rod-plane), and coaxial-cylinder configurations (Figure 5.1a–c). The pair of assemblies shown in Figures 5.1a and 5.1b is evocative, respectively, of the canonical schemes of

Filamentary Ion Flow: Theory and Experiments, First Edition. Edited by Francesco Lattarulo and Vitantonio Amoruso.
© 2014 by The Institute of Electrical and Electronics Engineers, Inc. Published by 2014 John Wiley & Sons, Inc.

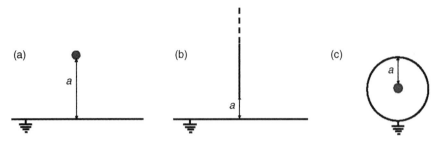

FIGURE 5.1   Illustrative electrode setups (collector grounded) adopted for corona and ion-flow investigations: (a) wire-plane (cross section) configuration, (b) rod-plane configuration, and (c) coaxial-cylinder (cross section) configuration.

an indefinitely long thin wire and a one-ended rod, the former being parallel and the latter orthogonal to an indefinitely extended planar collector a distance a apart. Both wire and rod are tacitly assumed as being round, while the tip geometry of the rod may remain unspecified (the excuse for this negligence will be cleared later in this chapter), provided that this conductor is slender enough. Equi- or multipotential montages, given by putting the individual examples of Figures 5.1a and 5.1b together, also happen to be adopted [see, for example, Sarma Maruvada (2000), Metwally (1996), and Jones (2006)]. For illustrative reasons, let us return to the previous referential pair of cases and allow the potential of the wire or rod to be raised to a value V exceeding the corona onset level V_T. As a consequence, an ionized thin region appears respectively distributed throughout or confined to the overstressed (in an electrostatic sense) termination. Once potently repelled into the drift region, the recently generated ions of same polarity as V spread up and slowly migrate toward the collector, thus taking a time of flight that typically ranges between tenths and unities of seconds, depending on V and the gap extension at hand. The electrode acting as both a collector and measurement surface for the impacting ions is generally grounded for safety reasons and for making ion-flow-related detections permissible and quite accurate. The ion-flow parameters of main interest are current density, charge density, and electric field, but the most indicative and easy to detect is the current density (see also below in this section). The instrumental investigation is restricted to the collector since any intruder violating the gap will often give rise to insuperable or untreatable disturbances. Basic advice about how the experienced investigator is going to perform laboratory activities pertaining to this argument can be found elsewhere [see, for example, Horenstein (1995) and (Maruvada, 2000, pp. 274–283)]. The amount of available experimental data concerning corona originated space-charge flows is large; however, there is a persistent need for exploring still overlooked or underestimated supplementary performances. These can emerge from special ion-flow field structures and can be carefully monitored while changing pivotal geometrical parameters one at a time. This is the rationale for including the present chapter in which unprecedented subsidiary electrode setups, even sometimes accomplished by unusual insulation, are introduced in order for the above object to be successfully attained. Owing to the informative character of some individual sources, a demanding experimental strategy has been adopted for making the investigation as complete as possible. To this end, a number of sources have been categorized in function of the injection system and methodically monitored. Therefore, the suspended electrode becomes a V-shaped wire, a two-conductor bundle, an inclined rod, and a partially sheathed wire, to name a few. It is tacitly understood that air is the filling gas at ambient, unless differently stated, pressure.

In this class of experiments involving corona activity, where a certain degree of instability is admissible, disturbing wanderings and fluctuations invariably affect detection of ion-flow parameters. To complicate matters further, short peaks, superimposed to larger distribution profiles of the quantity examined at the collector, need in some cases to be carefully perceived. In this regard, special care is to be taken to routinely remove electrophoretical depositions on the emitting surfaces to prevent, after prolonged operations, significant departures in the current distribution [see Amoruso and Lattarulo (1996)]. As a striking example, observe in Figure 5.2

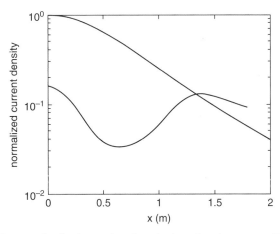

FIGURE 5.2 Current distribution at the plane in the wire-plane setup (3-mm-radius wire suspended at height $a = 70$ cm). Upper profile: Cleaned wire. Lower profile: Unclean wire (electrophoretically deposited ambient-air pollution). Applied voltage $V = -60$ kV. Normalization referred to the on-axis peak of the upper profile ($J_{c,max} = 7 \times 10^{-2}$ nA/cm^2).

the large depression in the current distribution appearing at short lateral distance from the central axis of the electrode geometry. This perturbation, caused by pollution because of nonuniform deposition all around a living wire above ground, disappears when the wire is clean (see curve comparison in the same figure). But this strongly recommended maintenance in itself is not enough. Under the above-mentioned critical expectations, the parameter deserving to be taken into account for the present kind of investigation must necessarily reduce, by elimination, to the current density alone (see later on). Direct detections of surface current distributions (from now on termed J_c-distributions, or something else, with subscript c indicating collector) appear in fact sensitive, rather stable and immune from intolerable disturbances. These qualities can, of course, especially be appreciated in laboratory, even though *in situ* measurements have extensively been performed anyway [see, for example, McKnight, Kotter, and Misakian (1983)]. Conversely, simultaneous electric field and space-charge density measurements [more detailed considerations can be found in Misakian (1981) and Misakian and McKnight (1987)] are, owing to different reasons, unpractical especially for localizing and recording potentially indicative second-order anomalies. Limitations to electric-field detection in the presence of space charge derives from the accessory space-charge-free field which is physically dependent on the electrode geometry but independent on space charge. The unpleasant consequence is that, especially in the cases of weak coronas, such a field component could be so important as to compromise the fieldmeter sensitivity to the presence of faint but meaningful concentrations of space charge. On the other hand, direct space-charge measurements are invariably performed by aspirator-type ion counters whose employment, however, is strongly warned against relatively high electric fields impinging on the inlet. These unpleasant circumstances exactly apply in the present class of experiments where ion flows are produced by high-voltage electrodes under corona. Ion counters not only suffer from the above limitations, but also make the measurement ultimately unfruitful because of wide (other than tedious) wanderings that concur to mask the fine structure of the distribution curves. In order to improve experimental perception of the above crucial anomalies by relatively short-dimensioned current probes, a wise practice consists in adopting electrode dimensions as large as possible, with the exception for cylindrically symmetric setups (the reason for this concession is explained later). In this respect, assiduous use is made of a $7.30 \times 2.30 \, \text{m}^2$ conducting plate as a ground collector placed beneath the emitter [see, as an example, Lattarulo and Di Lecce (1990)]. A nearly flush-mounted "Wilson plate-type" current probe is positioned on the geometric center of the grounded plane, thus collecting a very small fractional amount of the overall ion flow impacting ground. A set of easy-to-mount devices of this kind, an example of which is

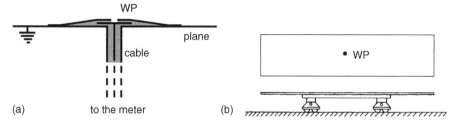

FIGURE 5.3 Laboratory plane. (a) Wilson-plate type (WP) current probe. (b) Motorized ground plane equipped with a centrally flush-mounted WP. Gray shading: Insulating compartment. Black tracing: Conducting compartment.

reported in Figure 5.3a, is constructed and made available since their form and reduced dimensions require to be assigned each time as a function of the configuration of the ion flow under examination. A cable connects the probe to a picoammeter equipped with a data acquisition system, the measuring complex being housed on a suitable shelf underneath the collecting plate. This arrangement ensures ion-flow perturbations and electromagnetic incompatibilities to be effectively circumvented. The grounded experimental apparatus is deposited on and carried by a motorized structure, in turn mounted on the laboratory floor rails as reported in Figure 5.3b. The rails assume the function of a guide for calibrated and slow translational motions. Thereby, planar J_c-distributions can be recorded during longitudinal explorations, often accomplished by local repetitions. These are performed by impressing minute back-and-forth shifts to the mobile plane in the neighborhood of specific locations on the explored track. Such a wise practice is extensively adopted here since those structures of the current density curves suspected to be meaningful are often rather confined and run the risk of being overlooked, irrespective of the detector sensitivity. Of course, it is mandatory to give the investigator the benefit of foresight in dealing with this class of experimental problems. The suspended emitter-bearing frame can remain motionless during each complete excursion or be simultaneously subject to rigid shifts transversal to the plate centerline. With reference to the latter option, detections throughout the collector surface become permissible as a result of a refined composition of mutually orthogonal motions, simultaneously impressed to the suspending system and probe-carrying collector. An equivalent envisioned picture of the above practice is that of a probe free to move, relative to a motionless electrode setup, over the wide planar region where the current fingerprint is permanently impressed. With reference to the coaxial arrangement of Figure 5.1c, use is extensively made of a commercially available airtight vessel, accomplished with gas-taps and a pressure gauge, in which the electrode system is mounted. The inner surface of the outer cylindrical collector, of diameter $2a = 97\,\text{mm}$ (the radius of the inner electrode is comparatively negligible), is entirely covered by an insulated array of adjacent strip-like Wilson plates. By being longitudinally extended throughout the available coaxial geometry, each elemental strip can capture large enough current that in this case short-dimensioned assemblies can be adopted without unpleasant consequences for the measurement quality. The plates are connected to a multiplexer enabled to select single probes consecutively and ultimately give discrete J_c-reconstructions. The vessel permits control of the inside gas pressure to an extent suitable enough to these applications. However, for an appropriate exploitation of this facility, care should be taken in assigning the applied voltage in that both the corona inception and gap breakdown level significantly depend on pressure.

Only a reasoned and restricted selection of databases, cumulatively forming a vast legacy of evidence after long-lasting (about three decades) activity by these investigators in their department high-voltage laboratory, has been aggregated, in this case compared to that illustrated in Figure 5.1, and then organically presented. In some cases, preference has been expressed for the result given with positive voltage. Under such conditions, data uncertainty due to corona instability holds within 5%, a limit value here systematically respected for ensuring good-quality results. The final part of the chapter is devoted to resolve a well-posed inverse source problem involving remotely detected observables. Specifically, corona source features are reconstructed starting from J_c-distributions previously measured at the collector. The ultimate intent is that of providing a method for appropriately assigning boundary conditions in terms of J_e-distributions (subscript e evokes the emitter) and, hence, general forward problems to be unambiguously and safely resolved. In this context, an eloquent discovery, although restricted to a set of ion flows showing bilateral or axial-symmetric symmetry, consists in the fact that the mutually different J_e-distribution laws so given can be unified according to a simple cosine law with index 2.

5.2 V-SHAPED WIRE ABOVE PLANE

A collection of J_c-distributions relevant to the case of a V-shaped long thin wire, raised to a voltage V exceeding the corona onset level, is displayed here by using different graphical descriptors. The grounded collector is the planar system previously introduced in Section 5.1. As shown in Figure 5.4, the source vertex, which is rounded-off rather than sharp owing to the finite cross-sectional dimensions of the wire (diameter 1 mm), points vertically toward and is suspended at height a above the ground plane. Both segments of the pair forming the suspended V-shaped emitter are inclined at the same

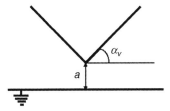

FIGURE 5.4   V-shaped wire-plane setup with generic inclination angle α_v. Specialized cases: $\alpha_v = 0°$ and 90°, respectively termed wire-plane (Figure 5.1a) and U-shaped wire-plane (with joined twin conductors, Figure 5.5) setups.

angle α_v with respect to the horizontal. The given symmetry in the source configuration is the prerequisite for reducing the illustrative figures, when required, to portions of the entire excursion without loss of information. The set of J_c-distributions monitored as α_v changes includes the important limit cases of $\alpha_v = 0°$ and 90°, respectively representative of the canonical wire-plane (see Figure 5.1a), and the suggestive but unfamiliar—hence, specifically termed—U-shaped wire examples. Note in Figure 5.5 that the twin segments of the adopted thin wire are joined as to form a vertical and compact two-conductor bundle, while the lowermost point of the U-shaped terminal is elevated above ground at height a. In spite of the resemblance of the arrangement of Figure 5.1b with that of Figure 5.5, the observed departures in the relevant J_c-distributions deserve, as will be appreciated later, careful discussion under a different heading. The suspended V-shaped wire can be rigidly translated, according to the description reported in Section 5.1, in the direction orthogonal to the motion of the planar collector. The current distribution is expressed in Figure 5.6a–r in the form of a set of 3D images from a (x, d_x, J_c) coordinate system centered on and oriented along the main centerlines of the grounded plate (in the figures, subscript x for the transversal coordinate d_x drops out for notational simplicity). Angles α_v and voltages V of both polarities are reported in the captions for each selected example, while an unchanged value of a, set equal to 30 cm, is to be

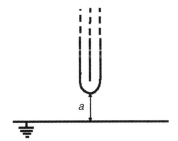

FIGURE 5.5   See Figure 5.4, $\alpha_v = 90°$ (not to scale).

tacitly understood without exception. Because of the given symmetries, the current "fingerprints" from the limit cases of $\alpha_v = 0°$ and 90° are conveniently reproduced in the form of J_c-curves and discussed separately (see later). For the sake of precision, the bilateral symmetry arising for $\alpha_v = 0°$ is to be restrictively intended for combination with a translational symmetry, thus in the sense that all the ion trajectories lie on crosscutting planes. On the other hand, the axial symmetry consequent to the condition $\alpha_v = 90°$ unambiguously evokes ion trajectories all lying on planes of a sheaf, the referential axis of which is the vertical through the origin of the coordinate system (central axis).

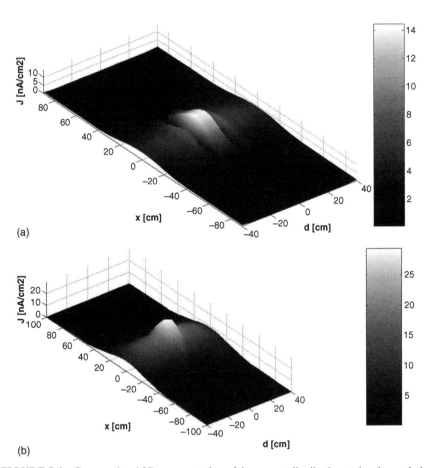

FIGURE 5.6   Gray-gradated 3D representation of the current distribution at the plane relative to the general case of Figure 5.4. Wire: Diameter $d = 1$ mm; vertex suspension $a = 30$ cm. (a, b, c, d, e, f): $\alpha_v = 15°$; $V = +40$, +50, +60, −40, −50, −60 kV, respectively. (g, h, i, j, k, l): $\alpha_v = 30°$; $V = +40$, +50, +60, −40, −50, −60 kV, respectively; (m, n, o, p, q, r): $\alpha_v = 60°$; $V = +40$, +50, +60, −40, −50, −60 kV, respectively.

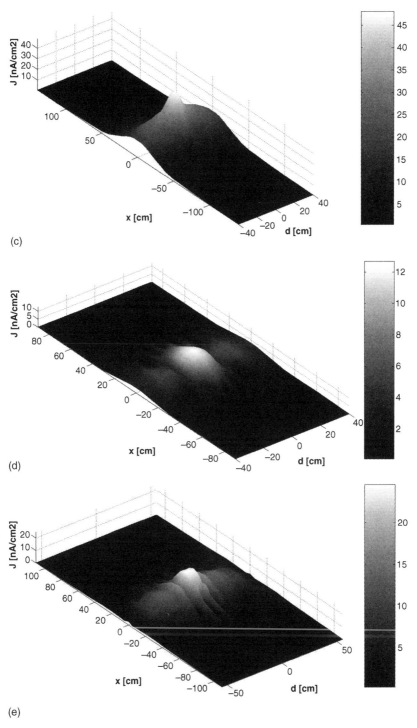

(c)

(d)

(e)

FIGURE 5.6 (*Continued*)

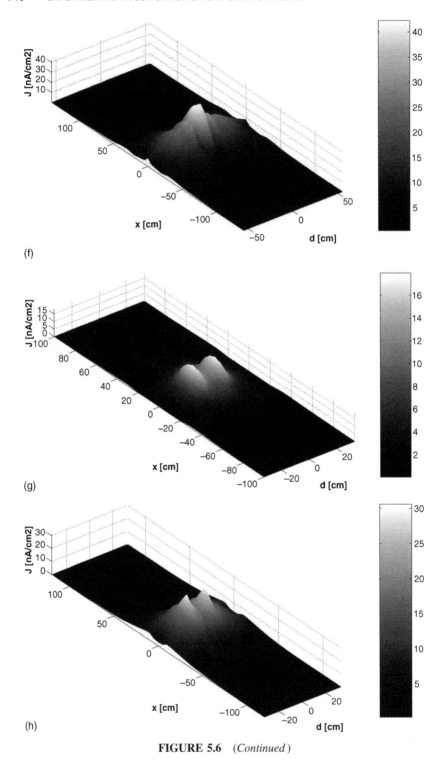

FIGURE 5.6   (*Continued*)

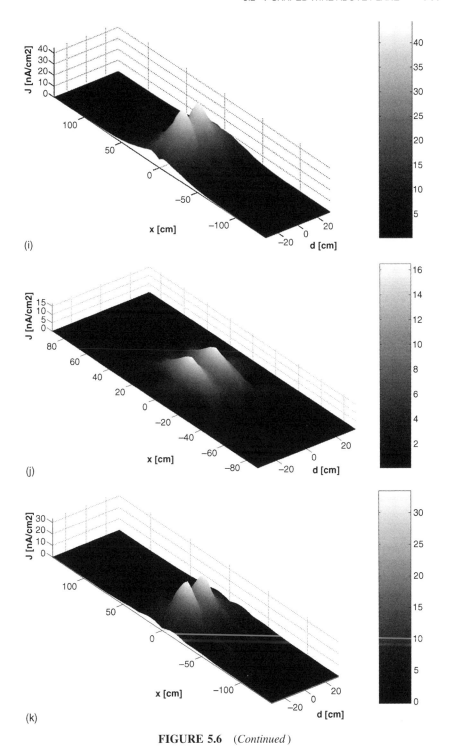

FIGURE 5.6 (*Continued*)

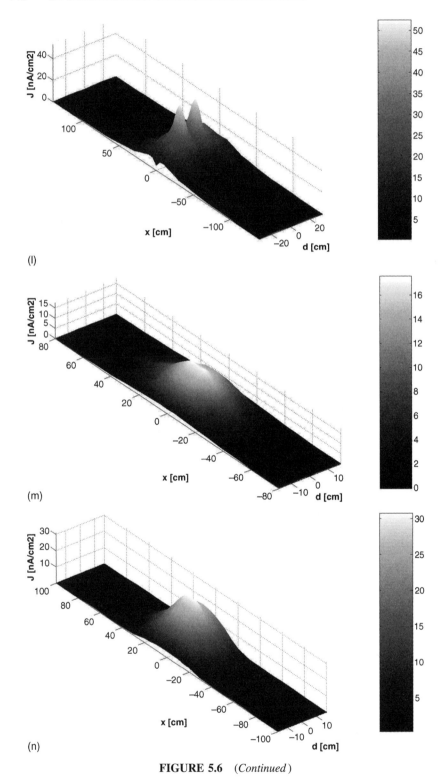

FIGURE 5.6 (*Continued*)

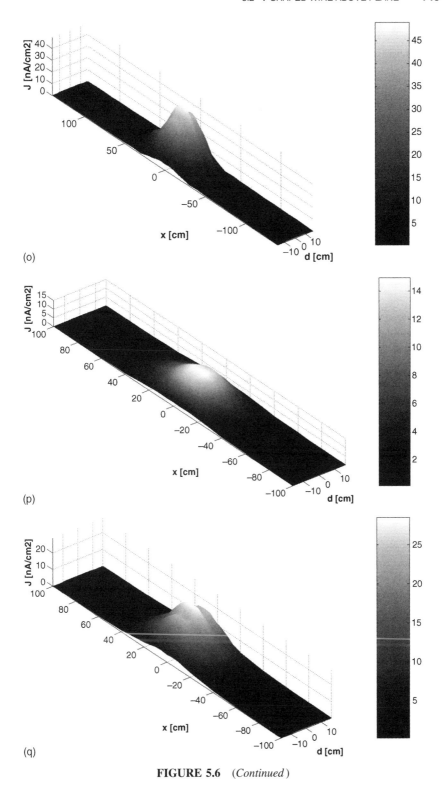

FIGURE 5.6   (*Continued*)

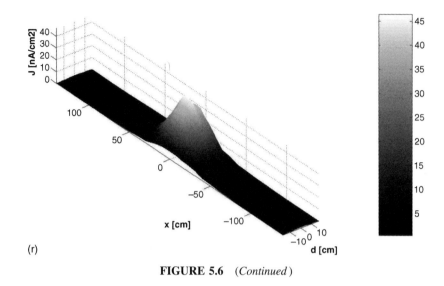

(r)

FIGURE 5.6   (*Continued*)

5.2.1 Main Observables

As α_v increases while V holds unchanged but is significantly larger than V_0, discernible and eloquent performances take place. In synthesis, the following events occur:

- The ion flow experiences a dramatic concentration just below the apex ($x = y = 0$) as soon as the initially straight wire ($\alpha_v = 0°$) is even slightly bent (Figures 5.6a and 5.6f).
- The concentration peak splits up into a pair of laterally symmetric and flattened humps as α_v increases further (Figures 5.6g and 5.6l). The most prominent point of each hump is positioned for $x = 0$ and $y = \pm a/4$, or thereabouts.
- The above pair of humps tend to be submerged by a centrally rising peak as α_v increases. Meanwhile, the overall ion flow is roughly prone to exhibit axial-symmetric configuration (see Figures 5.6m and 5.6r).
- The described performances are verified to be substantially irrespective of the emitter's elevation and voltage, the latter measured in magnitude and polarity. It is worth noting that even though α_v and V are adjusted as one wishes, the fixed height a seems to be the only parameter responsible for pinning down the lateral humps, until they become perceivable.

As promised, the above observations need to be supplemented by those given under the extreme circumstances dictated by $\alpha_v = 0°$ and $90°$. The pertaining features, expressed in the form of curves and related fittings, can be summarized as follows:

- The ion flow assumes, as doubtlessly expected, bilateral symmetry for $\alpha_v = 0°$ (referential wire-plane setup of Figure 5.1a), in which case the curves of the normalized J_c-distribution are fitted with a good approximation by a $\cos^n \theta$ law with power $n = 4$ (see Figure 5.7a). For the definition of the angle θ, see Section 6.8.1 and the related Figure 5.A.1. Small differences are detected when the polarity of V is reversed.
- The ion flow assumes, as confidently expected, axial symmetry for $\alpha_v = 90°$ (see Figure 5.5, U-shaped wire-plane setup with joined members of the given bundle), in which case the curves of the normalized

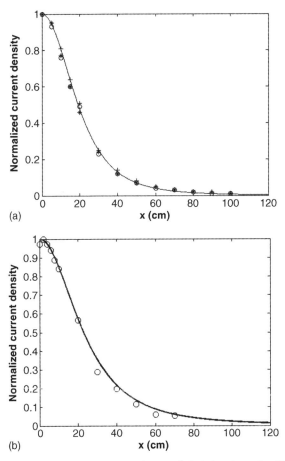

(a)

(b)

FIGURE 5.7 Experimental data and analytical $\cos^n \theta$ fitting (see also Figures 5.1a, 5.4, and 5.5). (a) $\alpha_v = 0°$, index $n = 4$; (b) $\alpha_v = 90°$, $n = 3$; (o) positive voltage; (+) negative voltage; $V = 60\,\text{kV}$ (absolute value). Wire: Diameter $d = 1$ mm; vertex suspension $a = 30$ cm.

J_c-distribution are fitted by a $\cos^n \theta$ law whose power n approaches 3 with a good approximation (Figure 5.7b). Major departures (not reported in the figure) are detected for this assembly when the polarity of V is reversed, maybe due to the larger mobility of negative ions. Reference could again be made to Figure 5.A.1, with the stipulation that the suspended referential point is, for reasons which will be clarified later, negligibly larger than the height a of Figure 5.5.

- The ion flow assumes, even for the above pair of extreme cases, normalized performances substantially irrespective of the emitter's elevation and voltage, with the specification that the peak related to the case of $\alpha_v = 90°$ is slightly affected by a central dimple when V largely exceeds the corona-inception level.

5.3   TWO-WIRE BUNDLE

The source type is here represented by twin equipotential wires whose mutual distance b is equal to (Figure 5.8a) or exceeding (Figure 5.8b) the diameter d. In the second case, an insulating spacer made of a pair of joined nylon threads of an assigned diameter, likewise composing a two-thread bundle, has been introduced. The pair of parallel wires are allowed to lie on the oppositely positioned reentrants of the insulating bundle, throughout, thus forming a two-wire bundle source with disjoined wires. The rotational symmetry so obtained for the overall system ensures that the centerlines of the pair of interpenetrated conducting and insulating bundles are perfectly superimposed. The collector is either the moving plate or the hollow cylinder, both introduced in Section 5.2. Such inactive electrodes are so widely spread apart that the Laplacian field is indistinguishable, except for a confined region just surrounding the active bundle, from one surrounding an equivalent single-wire distanced source of the same quantity a from the collector. Provided that the equivalence is rigorously applied to a per-unit capacitance in isolation (thus, for a tending to infinity), then the diameter of the equivalent one-wire is equal to $\pi d/2$. The same equality approximately applies to the above-mentioned

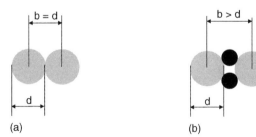

(a)   (b)

FIGURE 5.8   Two-wire bundle (cross section). (a) Joined wires. (b) Disjoined wires. Gray: Wire. Black: Non-scaled insulating spacer.

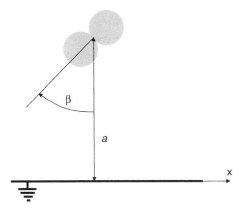

FIGURE 5.9 Two-wire bundle-plane setup (not to scale; cross section in Figure 5.8a).

practical cases because a is exceedingly larger than d. With reference to the example (treated here) of a U-shaped wire perpendicular to the grounded plate, the disjoining condition $b > d$ applies. Therefore, the given data and related discussion may be regarded as supplementary to those pertaining to the specialized geometrical conditions $b = d$ and $\alpha_v = 90°$. These have been met in the previous section with reference to a generic V-shaped wire. In hindsight, especially informative is the circumstance in which the pair of parallel segments of the wire respecting the disjoining condition $b > d$ are only a little distanced.

With reference to Figure 5.9, the normalized current density profiles at the planar collector are displayed in Figure 5.10 for the respective azimuthal positions

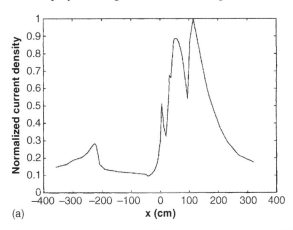

FIGURE 5.10 On-plane current density profiles relative to the setup of Figure 5.9. Wire: diameter $d = 1.4$ mm. Bundle: Suspension $a = 132$ cm; inclination $\beta = 45°$. Curves (a), (b), (c), and (d): Applied voltage $V = 52, 53.2, 60, 134$ kV (positive polarity), respectively. In particular, curve (d): Full and dashed (turned over) lines for hemicurves of the same distribution; dotted line for the cosine law with index $n = 4$ (wire-plane, Figure 5.1a). [Reproduced from Amoruso and Lattarulo (1998) with permission from *Journal of Electrostatics*.]

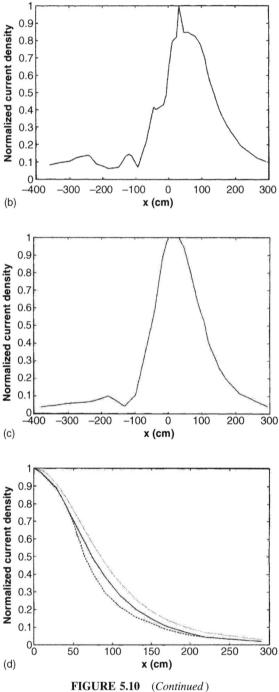

FIGURE 5.10   (*Continued*)

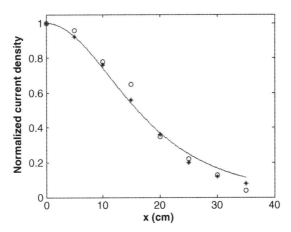

FIGURE 5.11 As Figure 5.10. Disjoined wires (Figure 5.8b). Wire: diameter $d = 0.6$ mm; bundle suspension $a = 25$ cm. Spacer: Thread diameter $d = 2$ mm ($b - d$ approaching d). Experimental data: ○ and + for $\beta = 90°$ and $0°$, respectively; $V = +50.4$ kV. Theoretical curve (full line): cosine law with index $n = 4$ (wire-plane case, Fig. 5.1a)). [Reproduced from Amoruso and Lattarulo (1998) with permission from *Journal of Electrostatics*.]

$\beta = 0°$, $45°$, and $90°$ and wire diameter $d = 1.4$ mm [see Amoruso and Lattarulo (1998)]. A similar case is represented in Figure 5.11, with the specification that d has been reduced to 0.6 mm, $b - d \cong d$, and, as a consequence (for reasons that will be clarified in the next subsection), only the extreme positions $\beta = 0°$ and $90°$ are taken into account. Incidentally, the diameter of each nylon thread of the insulating bundle, adopted as a spacer, is 2 mm.

Consistent with the coaxial arrangement of Figure 5.12 are the families of normalized profiles of Figures 5.13a–l, where evident reasons for symmetry

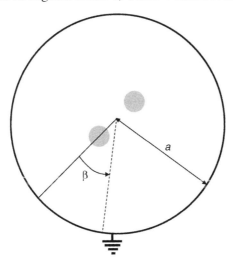

FIGURE 5.12 Coaxial setup (not to scale). Inner electrode: Two-wire bundle with spaced wires. Outer electrode: Grounded cylinder.

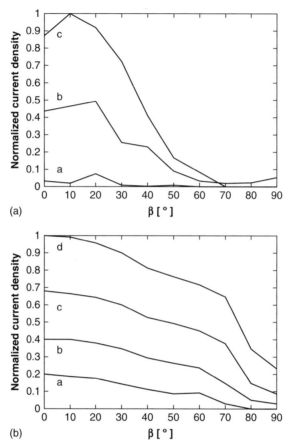

(a)

(b)

FIGURE 5.13   Experimental set of curves relative to the coaxial system of Figure 5.12 with $a = 48.5$ cm, $d = 0.32$ mm, $b = 0.98$ mm. Subset (a): Pressure 100 kPa; curves a, b, c with referential $J_{c,max} = 37.35$ nA/cm^2 and parameter $V = 13.0$, 13.5, 14.0 kV (respectively, positive polarity). Subset (b): Same pressure and $J_{c,max} = 1.11$ μA/cm^2; curves a, b, c, d with positive parameter $V = 18.0$, 22.0, 26.0, 30.0 kV. Subset (c): Pressure 60 kPa; curves a, b, c with $J_{c,max} = 33.1$ nA/cm^2 and positive parameter $V = 8.6$, 9.0, 9.25 kV. Subset (d): Same pressure and $J_{c,max} = 1.08$ μA/cm^2; curves a, b, c, d with positive parameter $V = 18.0$, 19.0, 20.0, 21.0 kV. Subset (e): Pressure 20 kPa; curves a, b, c, d with $J_{c,max} = 52.4$ nA/cm^2 and positive parameter $V = 4.25$, 4.5, 4.75, 5.0 kV. Subset (f): Same pressure and $J_{c,max} = 715$ nA/cm^2; curves a, b, c, d with positive parameter $V = 7.0$, 8.0, 9.0, 10.0 kV. Subset (g): Pressure 100 kPa; curves a, b, c, d with referential $J_{c,max} = -22.46$ nA/cm^2 and parameter $V = -13.0$, -13.5, -14.0, -14.5 kV. Subset (h): Same pressure and $J_{c,max} = -1.13$ μA/cm^2; curves a, b, c, d, e with parameter $V = -20.0$, -24.0, -26.0, -28.0, -30.0 kV. Subset (i): Pressure 60 kPa; curves a, b, c, d with $J_{c,max} = -68.0$ nA/cm^2 and parameter $V = -9.0$, -9.5, -10.0, -10.5 kV. Subset (j): Same pressure and $J_{c,max} = -987$ nA/cm^2; curves a, b, c, d with parameter $V = -13.5$, -16.0, -18.0, -20.0 kV. Subset (k): Pressure 20 kPa; curves a, b, c, d with $J_{c,max} = -426.0$ nA/cm^2 and parameter $V = -4.0$, -9.0, -10.0, -11.0 kV. Subset (l): Same pressure and $J_{c,max} = -1.63$ μA/cm^2; curves a, b, c, d with parameter $V = -8.0$, -9.0, -10.0, -11.0 kV. [Reproduced from Amoruso and Lattarulo (1998) with permission from *Journal of Electrostatics.*]

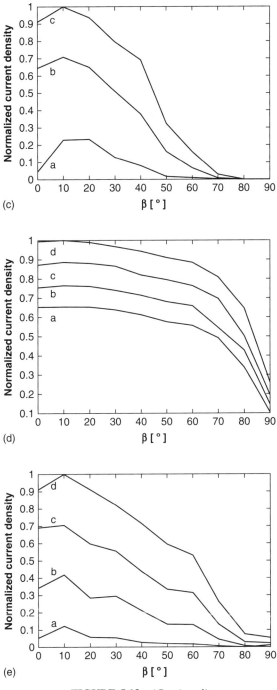

FIGURE 5.13 (*Continued*)

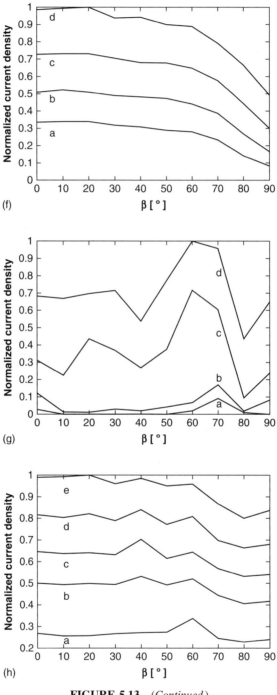

FIGURE 5.13   (*Continued*)

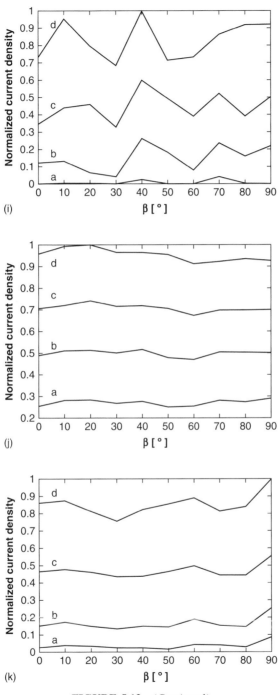

(i)

(j)

(k)

FIGURE 5.13   (*Continued*)

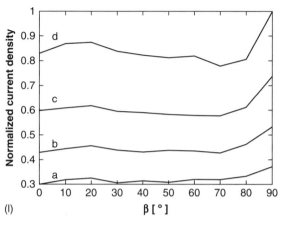

(l)

FIGURE 5.13 (*Continued*)

yield β only to span a quarter of the entire round angle without loss of intelligibility [see, again, Amoruso and Lattarulo (1998)]. The adopted spacing $b - d$ is only just enough to accentuate some differences in the given sets of curves for an easier interpretation of the physical mechanism under examination. Unlike the representation of Figure 5.10, given by using a unique Wilson plate linked up and moving with the motorized plane, the relative falling-off in quality of the broken-line one of Figure 5.12 of course depends on the discrete number of Wilson probes covering the outer cylindrical electrode. The numerical artifice of smoothing the connecting straight-line segmentation has been deliberately avoided. This is to stress, in a general sense, the superiority of the former detecting approach (here preferentially adopted as far as possible) that allows continuous curves to be directly supplied with good accuracy. The distinctive parameter of the family is gas pressure, while that of each curve composing a given family is the applied voltage. In both planar- and curved-collector setups, the referential current density adopted for the normalization is the highest peak J_M methodically detected along each profile.

The curves drawn in Figure 5.14, which include, in particular, the $\cos^3 \theta$ law curve for comparative purposes, are representative of the transversal U-shaped wire-plane geometry. Note that θ is for the semi-vertical cone angle of discharge (see also Section 5.8.1). The bundle members are here disjoined (Figure 5.8b) of a quantity $b - d$ approaching the wire diameter d.

5.3.1 Main Observables

The most striking performance monitored at increasing voltage for each case study regards the existence of an asymptotic ion-current distribution indistinguishable from that of a specified corresponding emitter of simpler form. This fundamental aptitude is invariably observed when use is made of

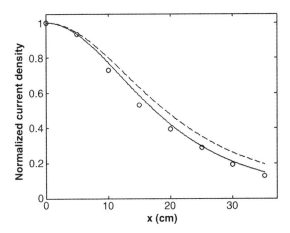

FIGURE 5.14 U-shaped wire-plane setup with disjoined twin members of the bundle (*b–d* approaching *d*, Figure 5.8b). Circles: Experimental data (averaged in a 40- to 60-kV voltage range, negative polarity); theoretical curves: cosine law with $n = 3.5$ (full line) and 3 (dashed line). Wire: $d = 1$ mm; $a = 25$ cm.

joined/disjoined wires and parallel/transversal planar or coaxial collectors. In particular:

- With reference to a horizontally stretched bundle with tilt angle $\beta = 45°$ (Figure 5.9), the on-plane current distribution (Figure 5.10) is subject to asymmetry. This expected feature, especially perceivable at lower voltages, results in a lateral shifting of the uppermost peak. Nevertheless, the peak is surprisingly displaced into the domain region, below the bundle, where the crossing Laplacian fluxlines spring up from the less stressed side of that source. As the voltage increases, the overall distribution is clearly prone to assume an on-axis bilateral symmetry governed by a cosine law with power 4. Such an asymptotic result is distinctive of the canonical wire-plane corona (single wire suspended).

- The above final considerations also apply (with the exception of neglecting those involving asymmetries) when the bundle is oriented with the limit angles $\beta = 0°$ and 90° (see Figure 5.11). However, comparing Figures 5.10 and 5.11 makes us realize that the slighter the bundle wire, the faster the monitored performance as a function of the applied voltage. In other words, the asymptotic distribution appears somewhat fully accomplished, even at relatively moderate voltages, provided that the wires are very thin. In this case, the possible presence of a narrow insulating spacer insignificantly influences the above behavior.

- The J_c-distribution (Figure 5.13) at the curved collector in the coaxial-cylinder geometry (Figure 5.12) tends to obey, especially under specified conditions, a $\cos^n \theta$ law with exponent $n = 1$.

- A comparison between Figures 5.7b and 5.14, regarding the case of an active U-shaped wire mounted transversally above a planar collector, sheds light on the first-order condition to be respected for a good fitting of experimental data with a $\cos^n \theta$ law with exponent $n = 3$. Substantially, the bundle needs to be formed by perfectly adjoining the two segments of the U-shaped wire $(b = d)$. The departure of the above law from experimental data starts being perceived as a nonzero gap $b - d$ between the two segments (Figure 5.8b) and is comparable with the wire radius, especially when the corona polarity is positive. For the sake of completeness, keep in mind that the above departure is quite unperceived, of course within unspecified limits for the gap $b - d$ and for $n = 4$, when a long bundle is horizontally stretched and revolved around its centerline above ground.

5.4 INCLINED ROD

The novelty introduced here into the rigidly archetypal rod-plane geometry consists of also imposing nonorthogonal orientations to the suspended rod. The corresponding ideal model identifies with a one-ended indefinite rod lying, along a given radial direction γ, on the vertical plane p cutting across the grounded plate (Figure 5.15). Similar to the case study treated in Section 5.2, the suspended rod can run, transversally to the motion of the grounded plate, over the p-plane. The current distribution on the planar collector is monitored for changing angular deviation $0° \leq \gamma \leq 90°$ of the

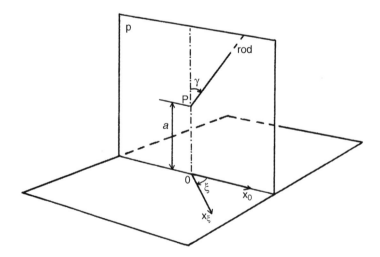

FIGURE 5.15 Geometrical references for the case of an inclined-rod above ground.

rod from the vertical (dash–dotted line), whereas the endpoint P is elevated at a fixed height a. The adopted physical emitter is a thin wire of variable radius tied in a slanting and rectilinear disposition by a light insulating support. This is essentially composed of a parallel array of nylon threads, coplanar with and transversal to the conducting wire, whose stretching peripheral frame is left free to rotate at the angle γ on the vertical plane p. Preliminary tests performed by using different supporting contrivances proved that corona-originated ion flows impacting the ground remain quite unperturbed. Rather, use has been made of a 10-m-long rod in order for the auxiliary corona-free junctions to the HVDC supply to cause a negligible electrostatic influence in the rod-plane gap. The physical endpoint has been realized by a net truncation of the wire, even though this specification is often of negligible importance. In fact, reiterated investigations report slight differences in the given coronas and corona-originated ion flows when a truncated rod is replaced, say, with conically or hemispherically tipped ones. The raised physical property is distinctive of slender rods [see, for example, McLean and Ansari (1986)], a prerequisite generally satisfied for several reasons in practical applications. Figures 5.16 and 5.17 show a pair of profile families as being detected by rectilinear excursions aligned with the x_0-axis, thus just below a one-ended and motionless horizontal wire ($\gamma = 90°$) of different radii [see Amoruso and Lattarulo (2005)]. In reality, it is the wire that is subject to refined collinear

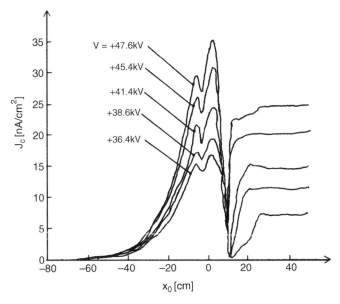

FIGURE 5.16 Experimental family of curves relative to the system of Figure 5.15 when $a = 27$ cm, $\xi = 0$, $\gamma = 90°$. Family parameter: Applied voltage, as reported. Wire diameter $d = 1$ mm.

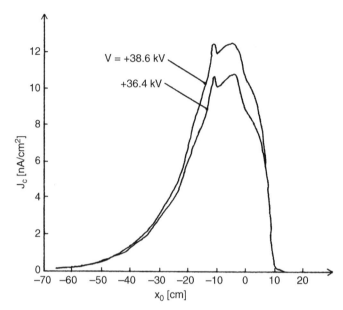

FIGURE 5.17   As Figure 5.16, with $d = 3$ mm. Curve resurgence cessation at the reported voltages.

shifts parallel to and distanced at a height a over a motionless ground plate. The curves reproduced in Figure 5.18 represent radial distributions, through the common origin 0, as being detected by surface excursions while, again, $\gamma = 90°$ holds unchanged. The family parameter is now the angular deviation ξ

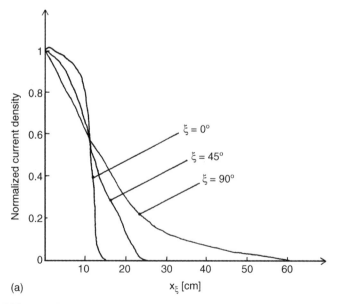

(a)

FIGURE 5.18   As Figure 5.16, with a selection of changing ξ as indicated in (a) and (b); $V = -47.6$ kV. Curve resurgence, only detected in (a) for $0 \leq \xi < 90°$, deliberately omitted.

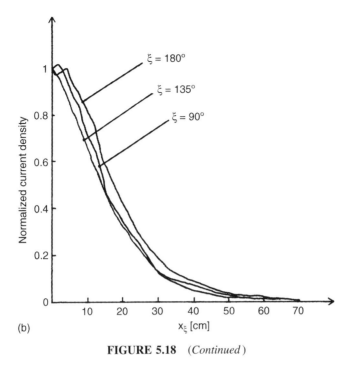

FIGURE 5.18 (*Continued*)

over the measuring plane of a given x_ξ-axis from the x_0-axis (Figure 5.15). Of course, the position coordinate x_ξ identifies with positive or negative values of x_0 when $\xi = 0°$ or $180°$, respectively. Instead, the data monitored along the x_0-axis (measuring plane motionless) when the wire is progressively brought in a vertical fashion are reported in Figure 5.19. Here, a set of centrally positioned curves as γ changes from $90°$ (horizontal wire) to $0°$ (vertical wire) are displayed.

5.4.1 Main Observables

These could be summarized as follows:

- Owing to the longitudinal extension of the profile set of Figure 5.16, it is easy to realize that each assigned voltage V (parameter of the curve family) exceeds the pair of corona onset levels relative to the tip (point P) and lateral surface of the emitter. Therefore, the x-abscissa assumes a positive or negative sign according to whether it is running along the wire projection or its external prolongation on the left-hand side of the origin 0 (identified, of course, with the on-plane projection of P). It is worth noting that each profile of the set appears decomposed into two separate compartments because of an inner abrupt J_c-collapse to zero invariably detected for $x_0 = 10$ cm, or thereabouts.

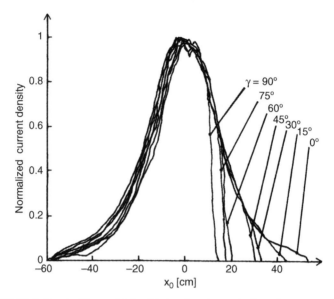

FIGURE 5.19 As Figure 5.16, with $\xi = 0$ and γ as indicated; $V = -47.6\,\text{kV}$.

- When the wire radius is 1.5 mm, the lower pair of the same voltages applied in the previous experiment are now above the corona onset level relative to the point but below one relative to the lateral surface of the rod. Therefore, the corona activity is only concentrated at the tip, with a subsequent cessation of current resurgence, beyond the collapse, underneath the wire (Figure 5.17).

- A qualitative comparison of Figures 5.16 and 5.17 offers convincing arguments to infer that, irrespective of voltage magnitude (and polarity, as additionally verified), the distribution law only ascribed to the terminal corona remains unchanged, regardless of the possible additional activity propagated along the rod. As an immediate implication, the discernible pair of ion flows whose sources are concentrated at the tip of, or disseminated along, the rod practically do not seem to be influencing each other. In particular:

- The J_c-distribution external to the rod projection (negative-x_0 range) obeys (until it vanishes) the Warburg $\cos^n \theta$ law with index n approaching 5 (WL). This fitting successfully includes the left peak of the asymmetrical dimple. On the other hand:

- WL fails along any radial direction except for the one aligned with the wire projection. Figure 5.18 offers an opportunity for this restriction to be appreciated since the curves display a dramatic violation of WL until ξ

departs from 180° and assume unspecified shapes, still different from WL, when ξ vanishes (just below the wire).

- Figure 5.18a shows that the current density abruptly collapses to zero for values of x_ξ progressively increasing with the angular deviation ξ until $0 \le \xi < 90°$. Under such conditions, the curve collapse is a little further followed by resumption (a detail not reported for the above figure), to appear as clear as possible. However, simple geometrical considerations lead one to expect that the impacting ion flow is as if it "leaves" a discontinuous "mark" on the ground plate because of a narrow zero-current band orthogonally crossing the x_0-axis. An example of the current fingerprint is reported elsewhere [see Figure 1.12 in Lattarulo and Amoruso (2007)].

- When the wire inclination is progressively reduced with respect to the vertical, the J_c-profiles underneath (positive side of the x_0-axis, Figure 5.19) can be roughly split into a pair of discernible curve segments in succession. These, imaginatively termed retaining- and collapse-range segments, exhibit a variable extension in the complementary sense that the former lengthens, at the expense of the latter, as γ decreases from 90° to 0°. Additionally, the retaining-range segment happens to be fitted by WL, namely, to be bilaterally symmetrical with the corresponding curve portion which lies on the negative side of the x-axis. Observing Figure 5.19 leads us to be persuaded that responsibility for the scanty superposition of the individual curves (of course for the portion of them obeying WL on both positive and negative sides of the x-axis), throughout, lies with the minute features differentiating the single profiles of the central dimple.

- As γ vanishes, the dimple-shaped profile attains an axial-symmetric central minimum, whereas the retaining-range curve segment is disposed to entirely cover the bell-shaped WL profile.

- For sake of precision, the above trend is often unaccomplished because of a remote cutoff perceived even under the familiar condition of γ = 0 (standing rod). In other words, when γ decreases to zero, the dramatic J_c-collapse depicted in Figure 5.19 appears so progressively distanced from the origin 0 and shortened as to be hardly detectable all around the rim of the final distribution. This is to stress that mounting the rod perpendicularly to the plane, as usual, unfortunately means to be worst placed to investigate some leading ion-flow performances.

- Under the above circumstances, no J_c-resurgence is detected beyond the cutoff even though the voltage magnitude is large enough to even cause ionization throughout the rod. As substantiated by experiment (Figure 5.19), the related current impacting the ground becomes unperceivable, irrespective of the applied voltage, because it is invariably subject to an excess of dilution.

- Reiterated tests prove that when thin rods are especially involved, the remote cutoff occurs for $x = x_M = a \tan^{-1} \theta_M$, the half-cone angle θ_M, with the cone apex on the tip, being nearly equal to 60°. This property substantially holds regardless of the assembly size and applied voltage, in magnitude and polarity.

- Altogether, WL fruitfully applies to the canonical rod-plane corona with the stipulation of passing some deficiencies corresponding to the presence of a central dimple and a final cutoff. Therefore, the fine structure of planar current distributions can be hardly reproduced simultaneously by using analytically manageable fitting formulas. Eloquent attempts are those previously made to reproduce the remote cutoff by Eq. (2.33) and, alternatively, the central dimple by Eq. (2.37). Indeed, such faint attributes are indicative of prominent phenomena, well deserving investigation, that are bound to remain submerged when the electrode orthogonality is, as emphasized before, invariably safeguarded. The reason why WL applies to the physical rod-plane case, although such a closed form is rigorously verified only for the unphysical point source-plane model, will be imparted in more general terms later on.

5.5   PARTIALLY COVERED WIRE

A wire enveloped in a longitudinally slotted insulating sheath is the special injector considered here (Figure 5.20) [a similar system is found in

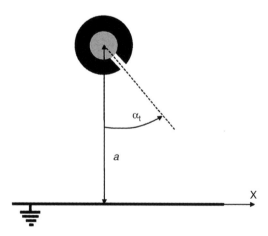

FIGURE 5.20   Partially covered wire-plane setup (not to scale). Gray shading up: Injecting wire. Black shading: Longitudinally slotted insulating sheet; α_t: slot inclination angle.

Amoruso and Lattarulo (1996)]. Once the wire is lodged in the round cavity
of a slotted dielectric sheet and raised to a given potential, complex corona
activity could take place all around the system. However, assigning both
sheet thickness and applied voltage carefully allows prominent ionization to
be confined, and narrow ion injection to take place, exactly where the slot is
positioned. Such a partial injector is horizontally stretched by corona-free
end arms, in turn connected to a bilateral suspending structure. An
important detail accomplishing the aerial system consists of a pair of
threaded elements interposed between the opposite ends of the injector
and the final stretching arms. The threaded elements are introduced for the
suspended injector to rotate around its centerline, namely, for the narrow
source to tilt with a given angle α_t, throughout. Setting $\alpha_t = 0°$ means that
the slot of the dielectric sheath has been brought to a position facing the
ground perpendicularly. Once it is preset and measured, the generic
orientation needs to be checked at regular intervals for accidental error
due to torsion of the described elongated device to be circumvented.
Therefore, the full extent of the system is methodically subjected to careful
goniometric tests in which the plane tangential to the sheath slot is assumed
as the reference. Samples of planar current profiles are reproduced in
Figure 5.21, where the relevant parameters of interest are reported. Each
curve is shown to be strongly dependent on the partial activity sustained at
a given voltage V and on the tilting angle α_t.

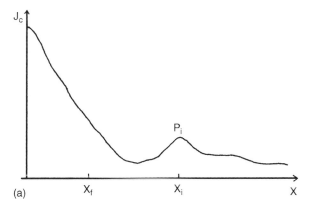

FIGURE 5.21 Experimental set of curves relative to the electrode system of Figure 5.20
suspended at $a = 70$ cm with wire diameter $d = 3$ mm. $J_c(x = 0)$ of the order of tenths to unities
of nA/cm^2, depending on V (ordinate scaling unimportant for this set of graphs). Subset (a), (b),
(c): $\alpha_t = 39°$ and $V = 70$, 90, 110 kV (positive); (d), (e), (f): $\alpha_t = 50°$, same sequence for V
(positive); (g), (h), (i): $\alpha_t = 39°$, same sequence for V (negative); (j), (k), (l): $\alpha_t = 50°$, same
sequence for V (negative).

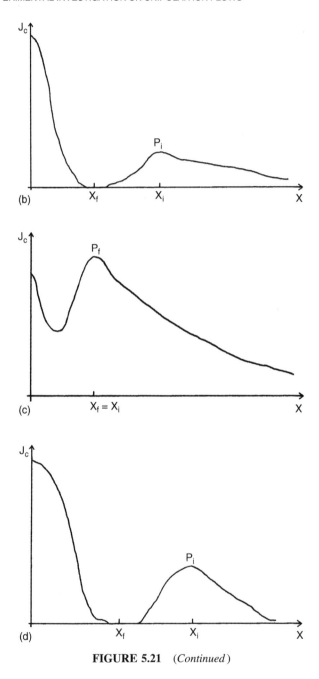

FIGURE 5.21   (*Continued*)

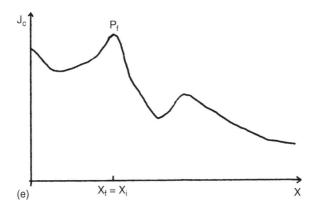

(e)

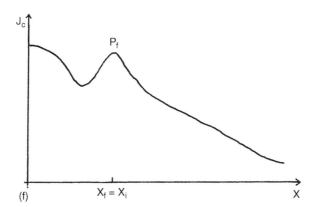

(f)

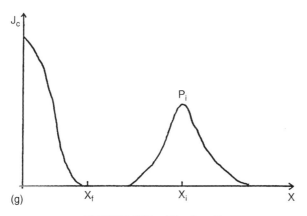

(g)

FIGURE 5.21 (*Continued*)

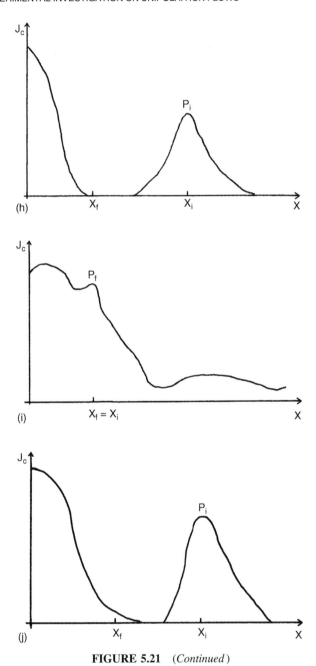

FIGURE 5.21   (*Continued*)

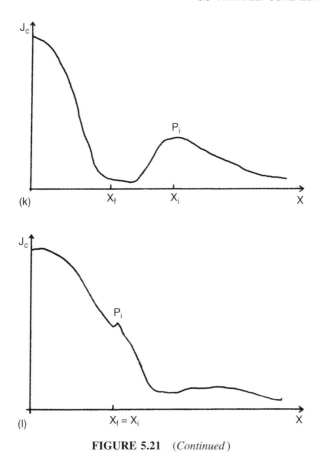

FIGURE 5.21   (*Continued*)

5.5.1 Main Observables

The previous experiments involving partially covered wires give further details to better see how the ion flow behaves in the illustrative wire-plane geometry and, *lato sensu*, in any given circumstance. Looking at the set of curves of Figure 5.21, whose J_c-profiles are to be taken as examples of repetitive performances given controlling both geometrical and electrical parameters, leads to realize the following:

- Only a fractional amount of the injected ion flow is able to convoy into the outwardly directed elemental Laplacian streamtube with injecting angle α_t from the narrow slot. In other words, the newborn ions are subject to voltage-dependent tangential spreadout, just beyond the slot, before being guided by the Laplacian pattern throughout the gap.
- The residual ion flow capable of escaping the initial spreadout is experimentally discernible in the form of a subsidiary peak, labeled

by P_f, in the J_c-profiles given at higher voltages. It is easy to verify that P_f is positioned on the destination coordinate $x = x_f = a \tan \theta$ with $\theta = \alpha_t/2$, thus in conformity with a Laplacian pattern in the typical wire-plane configuration. As a result of reiterated tests, a method is recommended to intercept P_f only recurring to experimental resources. It consists of increasing the applied voltage V starting from a quantity slightly exceeding the corona onset level and then watching that initially prominent peak P_i, remotely positioned at abscissa $x = x_i$, which is prone to shift back to the left of the figure and then to reach a position at abscissa $x = x_f < x_i$. This lower bound happens to be unsurpassable because the corresponding peak P_f persists in that position even when the voltage is increased further (see, in particular, Figures 5.21e and 5.21f, where the definitive abscissa $x = x_f$ is already reached at intermediate voltage). Of course, even the peak's form and height are strongly voltage-sensitive, as is the whole current distribution.

- At voltages far exceeding those involved in Figure 5.21, the rearranged ion spreadout causes P_f to be substantially smoothed, which implies that the actual J_c-profiles may be omitted being merely fitted by the already claimed $\cos^4$ law for uncovered wires.

5.6 POINTED-POLE SPHERE

An overwhelming difficulty arises in trying to ionize any thin gaseous layer surrounding a sphere or, at least, its overstressed cap in the sphere-plane electrode geometry. When the applied voltage is increased, in practice what happens is that an abrupt gap breakdown prevents the investigation on corona activity from being performed. This drawback could be circumvented by an artifice allowing the corona onset level to be sufficiently lowered with respect to the breakdown threshold. This has been made by attaching a sharp point to the sphere, exactly on the pole facing the distanced ground. The ultimate intent is that of reproducing an ion-flow pattern as being directly injected from a hypothetically active sphere above the plane. The point and upper sphere, whose equipotentiality is tacitly understood when a given voltage is applied to the sphere-plane assembly, play the actual combinative roles of concentrated ion source and extended surface for ion injection into the drift region, respectively. A detailed geometrical description of the asperity could be left unspecified; rather, the applied voltage needs to be carefully controlled, once the electrode geometry is assigned, owing to its influence on the morphology of the overall ionized region in the neighborhood of the point. To get to the heart of the problem, a superficial diffusion of ions just originated from the protruding point be established preliminarily over the surrounding cap of the suspended sphere in order for the ions, by that time injected into the drift region, to flow as being directly injected by the cap of the spherical electrode, namely, as if the latter was hypothetically smooth and active in the neighborhood of the overstressed pole. Indeed, when the applied voltage V is too low, or the sphere's electrostatic center-to-radius ratio a/r (see details in Section 5.6.1) approximately exceeds 60, the sharp asperity happens to practically behave as an ideal spot of charge in isolation. This occurs because grad n [see Eq. (4.B.3)] is too small for the surface diffusion to invade the upper cap of the sphere. An ion spreadout indistinguishable from one at the point of a rod is inferred since the detected current distribution at the plane obeys again the cosine law with power 5 ($\gamma = 0°$). On the other hand, at higher voltages and with a ratio a/r not exceeding the order of unities, the subsequent conical compression of the previous ion spreadout causes overconcentration of drifting ions around the revolution central axis. The planar current distribution in Figure 5.22 is striking evidence of the above description. Responsible for this second unwanted result is an exaggerated outward subtraction of ions from, thus at the expenses of a surface expansion of, the ionized cloud in the neighborhoods of the point source (see Appendix 4.B). In other words, the strong anisotropy of the actual ion source is explained because the intervals τ_D and τ have become comparable. All things considered, it is no use remarking again the amount of laboratory expertise required for a refined joint management of applied voltage and electrode geometry, with everything determined by minute changes in the remote detections at the collector.

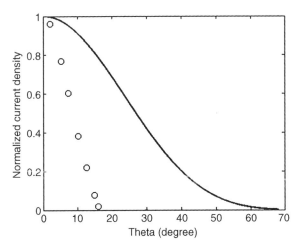

FIGURE 5.22 Pointed-pole sphere over ground plane. Full line: $\cos^n \theta$ law with $n = 6$ (expected theoretical curve, see Figure 5.23); Circle: experimental data for $a/r = 2$–6, nearly irrespective of the applied voltage in the range $V = 40$–65 kV (absolute voltage).

5.6.1 Main Observables

An appropriate use of the described injector, in substitution of an unsuccessful smoothed sphere, gives rise to further observables for a fruitful enrichment of those summarized up to now. The J_c-profiles regarding this case are obtained by using in a cumulative manner the entire amount of given databases. These are derived from tests involving a set of pointed-pole spheres of different radii, spanning a selected range of several centimeters. The common condition met for the construction of a given curve is that the experimental data, expressed in a normalized form, are those labeled with the same ratio a/r. Here, $a = \sqrt{h^2 - r^2}$ represents the electrostatic height of a sphere of radius r whose center is suspended at height $h > a$. In other words, once they are extrapolated from the spherical boundary inward, the Laplacian fluxlines converge to an off-center, low-lying point. Using the above definition, the general formula $\theta = \tan^{-1} x/a$ (see Figure 5.A.1) still holds and it can be seen that:

- The dissemination of experimental points is successfully fitted by a theoretical cosine law with power $n = 6$, irrespective of voltage magnitude and polarity (Figure 5.23), when $9 < a/r < 20$.
- With the same applied voltages, the fitting law holds similar even for a/r largely exceeding the above upper bound, with the stipulation that the index n reduces to 5 (Figure 5.24).
- According to the above results, the interrelated geometrical and electrical quantities contribute to confirm the intuitive expectation that the ion-flow pattern is substantially indistinguishable from that originated by a spot source of charges when the sphere and collector are far enough apart (in which case $n = 5$ and a approaches or exceeds h). Therefore, setting $n = 6$ is rather distinctive of a spherical source under the electrostatic influence of the plane (a appreciably less than h). For the sake of visual convenience, the implied fitting cosine laws with $n = 5$ and 6 are compared in Figure 5.25 with reference to an intermediate experimental circumstance causing the measured data to be distributed on complementary portions of both curves. Note the moderate mutual departure of the two curves, even though the Laplacian patterns guiding the respective fluxlines are described, as will be remarked in Appendix 5.A, by quite different analytical formulas.

5.7 STRAIGHT WEDGE

The edge of a dihedral conductor is positioned just in front of the ground plane at height a (Figure 5.26). This unfamiliar injector is accommodated in the

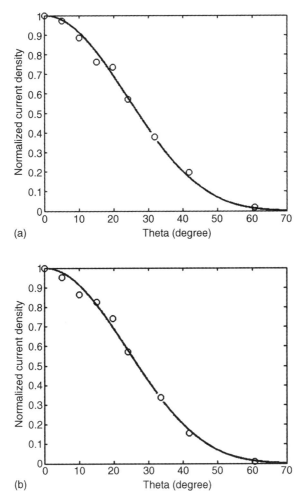

(a)

(b)

FIGURE 5.23 As Figure 5.22, $a/r = 9$–20. Average experimental data (circles) with negative (a) and positive (b) voltages.

present investigation especially because the informatory and versatile $\cos^n \theta$ law, governing previous current distributions at the plane, is proved to be preserved with $n = 2$ (this feature is also listed in Section 5.7.1 for the sake of convenience). However, differently from other cases under investigation, this meaningful formula is now applicable, provided that restrictive conditions are met. These will be clarified later and properly taken into account in monitoring the planar current density. The pair of intersecting planes forming the dihedron are made of conducting mesh in substitution of massive plates. The diameter of the mesh wires is 0.75 mm, and the straight edge of the meshed structure is as wide as the minor side of the moving plane. The latter is allowed to make broad excursions orthogonal to the upper edge since the planar projection of

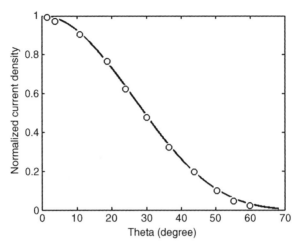

FIGURE 5.24   As Figure 5.22, $a/r = 78$; $n = 5$.

the wedge is of several meters, depending on the dihedral angle. The adopted mesh causes negligible electrostatic perturbations because these are confined within an imaginary pair of narrow layers sandwiching the discontinued material. The very motivation accounting for this practical solution is that two mutually conflicting requirements need to be somehow reconciled before carrying on the experiment. The wedge is demanded to be obtuse enough to allow the remotely positioned ion flow not to be significantly diluted. Under such circumstances, even the residual segment of the current profile, which is especially meaningful in the present investigation (see, again, later), is detectable. On the other hand, it is in principle desirable that the wedge be

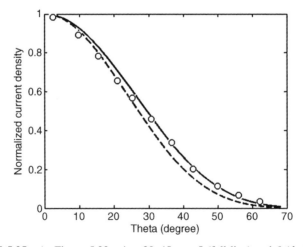

FIGURE 5.25   As Figure 5.22, $a/r = 20\text{--}45$; $n = 5$ (full line) and 6 (dashed line).

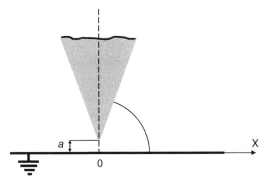

FIGURE 5.26 Illustrative wedge-plane setup with short gap *a*. Gray shading: Solid conductor. Black tracing: Generic fluxline approximately represented by a circle segment (thin); ground plane (thick).

acute enough, so that the task of triggering suitable corona activity on the rounded-off physical edge is made easier. Unfortunately, a wedge with an acute dihedral angle causes strong ion-flow decay transversal to the motion. As a consequence, the permissible excursion range on the measuring plane becomes too restricted to be fruitful. As specified above, a conducting mesh turns out to be the best practical solution because a suitable ion injection is ensured even when the edge presents a large dihedral angle. In other words, the distribution law of the current impacting ground is verified to qualitatively mimic the purely referential one envisaged to occur when using a massive wedge raised at an unpredictably high potential. As promised, it is urgent to account for the advice of shifting the on-plane detection, in the sense that the monitoring excursion is recommended to start from a given distanced position, with respect to the central axis, sideward. Bear in mind that the cosine law with exponent 2 evokes circular Laplacian fluxlines and that the prerequisite for this pattern to be reproduced, throughout, is that *a* reduces to a vanishingly small quantity. Unfortunately it is quite impractical to bring on the desired corona activity without causing a spark breakdown of the overstressed edge-plane short gap. On the other hand, the fluxlines are prone to depart from circular segments especially in a larger and larger central region of the Laplacian field as *a* is increased with the purpose of circumventing breakdown. A practical compromise is exactly attained by using a wedge-shaped mesh with an obtuse dihedral angle. When this injector is suitably elevated above ground, the laterally distanced and meaningful field subdomain—the one where drifting ions follow circular segments and ensure good perception of the decaying J_c-profile—broadens as needed. Further experimental specifications are supplied in the next subsection, where some results are selected and commented upon.

5.7.1 Main Observables

Even though the exploration range is rather restricted for the reasons discussed before, reiterated tests on this hard case study confirm the following:

- The planar current distribution in the wedge-plane seems to obey a cosine law with the stipulation of setting $n = 2$ (Figure 5.27a,b). This result is better substantiated in the case of negative ion flows.

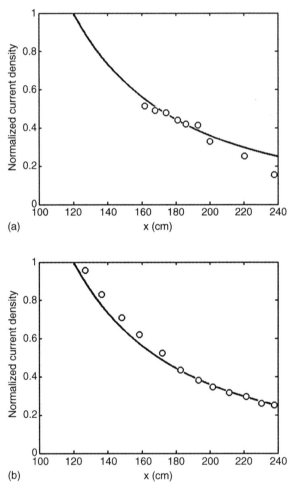

FIGURE 5.27 Experimental set of curves relative to the electrode system of Figure 5.26. The reported measuring excursion range ensures impacting currents guided by circular fluxlines in spite of the finite wedge elevation established in relation to the adopted voltages. Curves (a) and (b) differentiated for negative and positive polarity, respectively. Circles: Averaged values from databases given for changing applied voltage and elevation.

- The fitting under examination is fully discernible when the applied voltage substantially exceeds the onset level, irrespective of magnitude and polarity. In other words:
- In spite of the use of a discontinued material to deliberately increase the injecting performances of the suspended wedge, the monitored data at lower voltages are affected by such large wanderings and fluctuations as to frustrate any speculation on submerged meaningful features, if any.
- The profiles given at positive polarity often show a hardly perceptible remote cutoff (not reproduced here).

5.8 DISCUSSION

The remotely monitored (at the plane) collection of current distributions subject to different geometrical conditions give persuading arguments to infer that diffusion surrounding the source plays a crucial role in affecting the outer ion-flow morphology. In fact, diffusion can be remarkably responsible for a definitive reassessment of the boundary conditions in terms of the emitter surface's shaping and current distribution. The interested reader may look up the subject in Chapter 4, notably Appendix 4.B, where some useful tips, suitable to be implemented in available numerical codes treating ionization regions, are given. In fact, for a reliable evaluation of ion flows and related physical quantities, supplementary computational resources taking carefully into account this crucial phenomenon still represent an open problem. Some computational difficulties can be circumvented when special electrode geometries are adopted. In these favorable situations, an amenable analytical approach consistent with the theoretical framework of Chapter 4 and prone to inspire the physical interpretation of the results are made possible. Therefore, Section 5.8.1 has been conveniently devoted to such an informative analytical treatment. A number of the above observables are also discussed in a comprehensive form.

5.8.1 Supplementary Theoretical Analysis

The general theory examined in Chapter 4 (Section 4.4) provides a means for directly and unambiguously relating J_e at the emitter to the corresponding J_c to the collector. It is beneficial to reconsider Eqs. (5.A.3) and (5.A.4), skipped in Appendix 4.A, and write the rearranged formula as

$$\left(\frac{J}{J_0}\right)_e = \left(\frac{J}{J_0}\right)_c \left(\frac{E_{L,0}}{E_L}\right)_c \left(\frac{E_L}{E_{L,0}}\right)_e \tag{5.1}$$

Equation (5.1) reduces to

$$\left(\frac{J}{J_0}\right)_e = \left(\frac{J}{J_0}\right)_c \left(\frac{E_{L,0}}{E_L}\right)_c \tag{5.2}$$

provided that the equality $\left(E_L/E_{L,0}\right)_e = 1$ holds, namely, whenever the actual emitting surface is permitted to take a smoothed-out outer form, often different from that of the material conductor. As will be conclusively appreciated at the end of Appendix 5.A, responsibility for this realistic performance exactly lies with the overall ion cloud that is compartmented, in the simplified terms of the absolute figures described in Chapter 4, into the ionization and diffusion regions. The former is positioned in intimate contact with the material conductor and filled by an approximately zero-net space charge, while the latter becomes an enclosing narrow layer where a large amount of newborn unipolar ions are allowed to tangentially diffuse before being definitively repelled outward into a far wider drift region. The outermost surface between the above layer and the drift region properly represents that immaterial boundary whose configuration is of remarkable importance in determining the spatial ion-flow field. For applied voltages not exceedingly larger than the corona-onset level, the overall ion cloud is prone to assuming an asymptotic shape meeting the condition $\left(E_L/E_{L,0}\right)_e = 1$. This rounded-off boundary surface, which is the one where the recently generated ions are injected into the drift region, can feature an idealized substitute of the material electrode. Identifying the geometry of such a disfigured actual source is the prerequisite for finding a specialized relationship, according to Eq. (5.2), between J_e and J_c. The identification raised has been made discernible with the support of experimental data for the set of examples listed in the first column of Table 5.1 (also see Appendix 5.A).

TABLE 5.1   2D Case Studies

Geometry	Index n [Eq. (5.3)]	RSP[b]	Symmetry	Index p [Eq. (5.4)]
Wire-hollow cylinder	1	Wire center	Cylindrical	0
Wedge-plane	2	Edge	Bilateral	0
Line-plane[a]	3	U's lowermost point	Axial	1
Wire-plane	4	Wire center	Bilateral	2
Point-plane	5	Rod termination	Axial	3
Sphere-plane	6	Electrostatic center	Axial	4

[a]Idealized.
[b]Approximation.

With regard to the scheme of Figure 5.A.1, $\theta = \tan^{-1} x/a$ is the angle subtended at a suspended referential source point (RSP) by the planar segment of length x and a distance a away. Note that RSP is unambiguously determined for each source configuration (Table 5.1: source, first column; RSP, third column). Except for the special case of a cylindrical symmetry, θ represents a semi-vertical cone or wedge angle of discharge, in accord with the injection's axial or bilateral symmetry, respectively (symmetry axis orthogonal to the plane; Table 5.1, fourth column). Therefore, θ can label the individual Laplacian trajectory emerging with angle α from an injection point IP whose location depends on the injector geometry. Again barring the exceptional coaxial-cylinder geometry giving rise to radial trajectories which hold straight throughout, IP differs from RSP unless $\theta = 0$. Of course, this difference is a consequence of the curvilinear Laplacian pattern guiding the constituent trajectories.

In the light of Eq. (5.2), a fruitful exercise consists of finding $(J/J_0)_e$ once the formulas

$$\left(\frac{J}{J_0}\right)_c = \cos^n \theta \tag{5.3}$$

$$\left(\frac{E_L}{E_{L,0}}\right)_c = \cos^p \theta \tag{5.4}$$

have been established. As experimentally [Eq. (5.3)] and analytically [Eq. (5.4)] verified for each examined case, the above pair of equations manifest a similar structure with exponents n and p equal to the numerical values reported in Table 5.1. Every source listed in the first column of this table is inherently physical or idealized, as the case may be. Before going on to find an expected cosine law for $(J/J_0)_e$, some further words are hereafter spent to accomplish the presentation of Table 5.1 and, therefore, better appreciate how Eqs. (5.3) and (5.4) come forward. Some detailed considerations pertaining to the numerical values listed in the second and fourth columns of Table 5.1 can be found in Appendix 5.A.

The familiar "thin-wire" meaning is tacitly understood wherever the simpler term "wire" happens to appear. This mention justifies the fact that the electrostatic center in the wire-plane geometry, rigorously representing RSP, has been replaced with the wire center without an appreciable loss of precision. As a result, the Laplacian pattern accounting for the equality $n = 4$ in the second column of Table 5.1 descends from an appropriate analysis involving a bipolar-coordinate system. Returning to the first column of Table 5.1, consider that the adopted U-shaped wire under a corona behaves in fact as a hyperboloid whose degenerated approximation retrogresses toward a nonuniformly emitting one-ended line above the plane. According to this ideal figure perpendicular to the plane (Table 5.1, first column), the Laplacian trajectories are a quarter of ellipses and RSP (Table 5.1, third column) is the line endpoint whose position rigorously represents the focus of the confocal hyperboloid equipotentials orthogonal to the likewise confocal elliptic flux-lines of the Laplacian substrate. This unphysical line and its termination (the focus of the above orthogonal pattern) are respectively argued to be (a) collinear with the longitudinal centerline and (b) nearly coincident with the lowermost point of the physical U-shaped wire. It is high time to realize that an indefinitely long wire (rod) and a one-ended two-wire bundle with a U-shaped termination represent two examples of an identical topological isomorphism. The practical explanation lies in the continuous bending of the initially endless configuration into the U-shaped one. The raised geometrical property is the prerequisite for expecting that the given two-wire bundle and the original wire—the latter intended as being degenerated into a filament of vanishing cross-sectional area (roughly speaking, as being an active rod, throughout, with an inactive tip)—are interpreted from a crosswise distant collector as indistinguishable ion-flow straightline sources. Instead, the activity of a physical rod is perceived by the same collector as a point spot of charge concentrated on the tip. This because the rod end now represents a topological transition, no matter whether it looks as a rounded-off, tapered, or truncated one. Therefore, going on in the presentation of Table 5.1, the logical consequence is twofold: RSP regarding the rod-plane case is approximately positioned on the rod termination (third column) with no regard for its detailed geometry; the Laplacian pattern guiding the ion trajectories is that pertaining to the customary case of a point charge above ground.

Instead, the real dimensions of a spherical electrode above the influencing plane are responsible for the rigorous positioning of RSP in the electrostatic center of the conducting sphere (third column).

At last, methodically rearranging Eqs. (5.2), (5.3), and (5.4) and introducing the appropriate numerical values into the pair of exponents (n, p) listed in Table 5.1 gives

$$\left(\frac{J}{J_0}\right)_e = \cos^2 \theta \tag{5.5}$$

The invariant character of Eq. (5.5) for the analytically tractable case studies listed in Table 5.1 perhaps represents the most striking issue of this investigation, and on that account it deserves careful consideration. Accordingly, the optional version of Eq. (5.5) [see Eq. (5.A.4)], directly expressed as a function of the injection angle α_i when the source is a wire, now appears to be especially informative. At first sight, it seems hard to grant that a nonuniform current distribution law, notably Eq. (5.A.4), becomes reconciled with the physical circumstance of a thin wire horizontally emitting at a large height above the ground plane. A hasty intuition is that the ion injection law around a wire subtracted from the electrostatic influence of the far plane must hold to be perfectly isotropic, not according to a cardioids law, as otherwise Equation (5.A.4) reads. Therefore, if the multichanneled ion-flow model of Chapter 4, on the one hand, claims the constancy of J_i and $\mathcal{S}$ for each elemental channel, on the other hand, the same model proves that it is the number density $(n)_e$ of the emerging channels which must differ around the emitter. Indeed, the former performance (J_i unchanging) is irrespective of the electrode configuration, while the latter [$(n)_e$ changing] is instead dependent on the overall, rather than local, structure of the Laplacian field. In other words, the very fact that the initially radial channel-guiding Laplacian fluxlines tend to become curvilinear toward and orthogonal to the plane is sufficient reason to expect, in light of the above model, that the channels are nonuniformly distributed around the cylindrical wire. Similar observations apply to the point-plane and, to some extent, to the sphere-plane geometry because even in these cases the ion injection starts radially from the respective RSP, but the electrostatic influence of the plane forces the trajectories to become bent.

5.9 GENERALIZATION ACCORDING TO INVARIANCE PRINCIPLES

The various nonuniformities expressed, in the interests of expediency, in Appendix 5.A by way of Eqs. (5.A.4), (5.A.8), (5.A.10), and (5.A.11) prove the special dependence of the current distribution at the emitter to the whole Laplacian field. This performance is formalized by Eq. (4.25), in turn connected to the key ratio dN/dN_0 through Eq. (4.A.11). Therefore, the fact that $(E_L/E_{L,0})_e = 1$ simultaneously applies to ionized fields with a curvilinear pattern cannot be put forward as an argument for likewise expecting the equality $(J/J_0)_e = 1$. Also, boundary conditions expressed as ratios of the kind $(J/J_0)_e$ and $(E_L/E_{L,0})_e$—the former referred to as given voltage V exceeding the corona onset V_0, the latter referring to a voltage that instead is not exceeding V_0—are both rather determined by the overall, rather than local, morphology of the actual Laplacian fields, albeit in different manners. These remarkably influence, in particular, the ratio $(J/J_0)_e$ even when diffusional effects just surrounding the emitting electrode are negligible, namely, when the Laplacian pattern is allowed to remain unchanged in both circumstances (see the above-mentioned wire-plane case). As far as the restricted list of examples of Table 5.1 is concerned, the set of distribution laws given by Eqs. (5.A.4), (5.A.8), (5.A.10), and (5.A.11) also happen to be surprisingly unified in the form of Eq. (5.5). This has been made permissible because the parameter labeling the injection point—that is, angle α or distance b, whatever the configuration of the electrode assembly may be—has been replaced with the corresponding angle of discharge θ. Apart from this geometrical contrivance, diffusion certainly plays a physical key role in making Eq. (5.5) successful.

Finally, it seems reasonable to take special advantage of the amount of evidence collected previously in order to see them in light of invariance principles applied to ion-flow fields. The very purpose of this reading is that of getting insight into the generalization of formulas originally found for 2D fields. Regarding this, the comments made can be succinctly summarized as follows:

- Equation (5.5) is interpreted as an invariant observable (i.e., a symmetry) under corona changes.
- The ratio dN/dN_0 discussed in Appendix 4.A successfully runs for a surrogate of the changing coronas because it behaves as a generator of configurations, namely, an operator of a symmetry transformation, preserving Eq. (5.5).
- A foremost property attributable to Eq. (5.5) is that it is unaffected by the curvilinear features of general ion-flow fields and is only dependent on a radial straightline pattern. This converges upon a referential source point (RSP) that needs to be specified in each individual case study.
- For each configuration, it is then permissible to ascribe to the straight line of length $a/\cos\theta$ (Figure 5.A.1) the function of the referential trajectory of least curvature, among others, with arclength $L > a/\cos\theta$ sharing the same stationary condition (Hertz's action principle). This kind of condition is consistent with any given Laplacian trajectory pattern and represents the prerequisite, according to Noether's theorem, for the symmetry of laws of the kind $(J(p)/J_0(p_0))_e$ (p is for α or b) to correspond exactly to the conservation law $(J/J_0)_e = \cos^2\theta$.
- Because the homeomorphism involving the ratio dN/dN_0 applies regardless of the assembled electrode system, the meta-character of the invariance principles which have come into play allows the above 2D issues to be extended to general 3D ion-flow fields without fear of symmetry violation.
- Equation (5.5) is expected to hold even though some tentative applications could apparently give rise to symmetry breaking. This might happen when positioning RSP becomes an arbitrary exercise leading to substantial error. What can be said briefly is that for several applications of practical interest, RSP approximately identifies with the more stressed location of the variously shaped material emitter. Therefore,

a becomes the referential distance of that point from the generic collector.

- The invariance expressed by Eq. (5.5) furnishes additional arguments in favor of a generalization for the asymptotic condition $\left(E_L/E_{L,0}\right)_e = 1$, *a priori* only assumed for the case studies of Table 5.1.

- In accordance with the above condition, there is a unified outer shaping of the ionization region, thus resulting in a smoothed (basically, spherical or cylindrical) injecting surface, for all coronas. As regards such a specific feature, simple geometrical considerations should lead us to become fully aware of the very confined extension of this injecting surface matched with the rest of the living conductor and, more than that, with the widespread drift region.

APPENDIX 5.A

Analytical and experimental data are combined here to ultimately provide additional arguments for a correct approach to the complex phenomenology under examination. As the reader will appreciate, the subjects listed in Table 5.1 (see first column) are explored following a different succession for explanatory reasons. According to this specification, it seems convenient to start from the wire-plane geometry by imposing current continuity to an elementary fluxtube. As a result, we have

$$J_c = J_e r_0 \frac{d\alpha_i}{dx} \tag{5.A.1}$$

whose permissible reformulation is

$$J_c = J_e r_0 \frac{2\cos^2\theta}{a} \tag{5.A2}$$

Equation (5.A.1) is derived from simple geometrical considerations involving the per-unit cross sections $dA_e = r_0 d\alpha_i$ and $dA_c = dx$ at both ends of the fluxtube and the injection angle α_i with respect to the vertical (see Figure 2.1). Equation (5.A.2) is given by analytically performing the derivative $d\alpha_i/dx = d\alpha_i/d\theta \cdot d\theta/dx = 2 \cdot (\cos^2\theta)/a$. Use has been made in this case of the equality $\alpha_i = 2\theta$, pertaining to a bipolar-coordinate system applied to the circular trajectories, also remembering the general relationship $x = a \tan\theta$ involving the semi-vertical wedge angle of discharge θ indicated in the geometrical construction of Fig. 5.A.1. In addition, an outer cylindrical injection of unspecified radius r_0 slightly distanced from the wire surface is

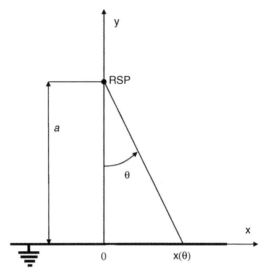

FIGURE 5.A.1   General scheme for a referential source point (RSP) above plane. Semi-vertical cone or wedge angle of discharge $\theta = \tan^{-1} x/y$; $y \geq a$ with a representing the elevation of the lowermost point of the active electrode. With reference to the examples listed in Table 5.1, $y = a$ but the theoretical line-plane case (physically simulated in laboratory by a U-shaped-plane setup elevated at height $y = a$).

assumed because no significant tangential diffusion is reasonably permissible all around an evenly, or approximately so, active wire in isolation. From Eq. (5.A.2), we obtain

$$\left(\frac{J}{J_0}\right)_c = \left(\frac{J}{J_0}\right)_e \cos^2 \theta \qquad (5.A.3)$$

so that Eq. (5.5) derives once the experimental law $(J/J_0)_c = \cos^4 \theta$ is taken into account. After remembering the key equality $\alpha_i = 2\theta$, Eq. (5.5) can alternatively assume the cardioid law

$$\left(\frac{J}{J_0}\right)_e = \frac{1 + \cos \alpha_i}{2} \qquad (5.A.4)$$

if preference is expressed for a polar form [see also Eq. (2.34)]. It is worth noting that the illustrated features of the curvilinear trajectories represent the prerequisite for the above injection law to be an anisotropic one. This is to stress, on the other hand, that it is no accident that the isotropic injection law $(J/J_0)_e = 1$ uniquely derives from imposing the angle equality $\alpha_i = \theta$ (coaxial arrangement determining straightline radial trajectories, throughout).

Also circular are the trajectories pertaining to a sphere-plane electrode assembly whose axial-symmetric configuration, to which a bi-spherical coordinate system rigorously applies, accounts for the discernible differences between the given formula

$$J_c = J_e r_0^2 \frac{\sin \alpha_i}{x} \frac{d\alpha_i}{dx} \tag{5.A.5}$$

and Eq. (5.A.1) (remember that the latter is instead relevant to a bilateral symmetry). Incidentally, a sphere of unspecified but very short radius r_0 has been introduced as a substitute to the true spherical injector whose radius is r_s and geometrical center's height is h_s. This replacement is admissible, provided that the geometrical center and potential of the short sphere are respectively lowered at height $h_0 = (a^2 + r_0^2)^{1/2} \cong a$ and raised at a value proportional to $ch^{-1}h_s/r_0/ch^{-1}h_s/r_s$. Here, $a = (h_s - r_s)^{1/2}$ identifies the height of the common electrostatic center. Briefly, Eq. (5.A.4) legitimately holds and Eq. (5.5) is again found because

$$\left(\frac{J}{J_0}\right)_c = \left(\frac{J}{J_0}\right)_e \cos^4 \theta \tag{5.A.6}$$

with $(J/J_0)_c = \cos^6 \theta$ provided by data fitting (as specified in Section 5.7.1, θ denotes now the semi-vertical cone angle of discharge).

Now turning our the attention to the point-plane case, Eq. (5.A.5) is still applicable to this axial-symmetric geometry with the specification that the tipped rod is ideally replaced by an isolated point spot of vanishing radius r_0 centered on the tip itself. It is worth noting that the limit represented by r_0 tending to zero forces the curvilinear trajectories to depart from the circular segments implied in the previous cases. In this regard, reference could be made to basic textbooks reporting the field map surrounding a point charge above a conducting plane. There is no difficulty in verifying in particular that $\cos \theta = (1 + \cos \alpha_i)/2$ and, in turn, Eq. (5.5) is again obtained because Eq. (5.A.5) becomes

$$\left(\frac{J}{J_0}\right)_c = \left(\frac{J}{J_0}\right)_e \cos^3 \theta \tag{5.A.7}$$

where $(J/J_0)_c = \cos^5 \theta$ is experimentally known. From the above θ–α function, it follows that

$$\left(\frac{J}{J_0}\right)_e = \left(\frac{1 + \cos \alpha_i}{2}\right)^2 \tag{5.A.8}$$

In a word, the presence of a cardioid law seems invariably evocative of trajectories represented by circular segments, the squared version of which takes into account the increased anisotropy of the axial-symmetric injection with respect to the bilaterally symmetrical circumstances leading to Eq. (5.A.4).

When the suspended source behaves as an ideal line (endpoint deprived of emission) orthogonal to the plane, then the substitute of Eqs. (5.A.2) and (5.A.5) becomes

$$J_c = J_e \frac{r_0}{x} \frac{dy}{dx} \tag{5.A.9}$$

Even now Eq. (5.5) is verified because $(J/J_0)_c = \cos^3 \theta$ is given by experiment, and the axial-symmetrically emitted trajectories are a quarter of the confocal ellipses (focus positioned on the line endpoint). Under such circumstances, we have $dy/dx = x/y = (x/a)\cos \theta$ and

$$\left(\frac{J}{J_0}\right)_e = \left(\frac{a}{b}\right)^2 \tag{5.A.10}$$

Here, the vertical coordinate y, equal to a plus the distance of the individual emission point from RSP, has been replaced with a more generic symbol b for notational convenience (discover below the reason for this substitution). Last, when the source is a wedge just facing a plane (vanishing a), the trajectories become concentric circular segments. As a consequence, the cross section of any elemental fluxtube holds constant and, in turn, the equality $J_c = J_e$ represents the trivial substitute of Eqs. (5.A.1), (5.A.5) and (5.A.9). Because $(J/J_0)_c = \cos^2 \theta$ is experimentally verified, Eq. (5.5) is confirmed without further manipulations. Importantly, Eq. (5.5) can be reformulated as follows:

$$\left(\frac{J}{J_0}\right)_e = \left(\frac{a}{a^2 + b^2}\right)^2 \cong \left(\frac{a}{b}\right)^2 \tag{5.A.11}$$

thus bearing some resemblance to Eq. (5.10) since now a tends to vanish and $x = b$.

REFERENCES

Abdel-Salam M. Calculating the effect of high temperatures on the onset voltages of negative discharges. *J. Phys. D: Appl. Phys.* 1976; **9**(12):L149–L154.

Abdel-Salam M, Al-Hamouz Z. Novel finite-element analysis of space-charge modified fields. *IEE Proc. -Sci. Meas. Tech.* 1994; **141**(5):369–378.

Abdel-Salam M, El-Mohandes MT, El-Kishky H. Electric field around parallel DC and multi-phase AC transmission lines. In: *Conference Record of the 1989 IEEE, October 1-5, 1989, San Diego, CA.* Industry Applications Society Annual Meeting, pp. 2014–2020.

Abdel-Salam M, Farghally M, Abdel-Sattar S. Finite element solution of monopolar corona equation. *IEEE Trans. Electr. Insul.* 1983; **EI-18**:110–119.

Abdel-Salam M, Stanek EK. On the calculation of breakdown voltages for uniform electric fields in compressed air and SF6. *IEEE Trans. Ind. Appl.* 1988; **24**(6):1025–1030.

Aboelsaad MM, Shafai L, Rashwan MM. Improved analytical method for computing unipolar DC corona losses. *IEE Proc. Pt. A* 1989a, **136**(1):33–40.

Aboelsaad MM, Shafai L, Rashwan MM. Numerical assessment of unipolar corona ionized field quantities using the finite element-method. *IEE Proc. Pt. A* 1989b, **136**(2):79–86.

Adamiak K. Simulation of corona in wire-duct electrostatic precipitator by means of boundary element method. *IEEE Trans. Ind. Appl.* 1994; **30**(2):381–386.

Filamentary Ion Flow: Theory and Experiments, First Edition. Edited by Francesco Lattarulo and Vitantonio Amoruso.
© 2014 by The Institute of Electrical and Electronics Engineers, Inc. Published by 2014 John Wiley & Sons, Inc.

Adamiak K. Numerical models in simulating wire-plate electrostatic precipitators: A review. *J. Electrostat.* 2013; **71**(4):673–680.

Adamiak K, Atten P. Simulation of corona discharge in point-plane configuration. *J. Electrostat.* 2004; **61**(2):85–98.

Allibone TE, Saunderson JC. Corona at very high direct voltages. VI corona in rod/plane gaps. In: *Proceedings of the VI International Symposium on High Voltage Engineering (ISH'89),* August 28–September 1, 1989, New Orleans, LA, Paper 22.02.

Allibone TE, Jones JE, Saunderson JC, Taplamacioglu MC, Waters RT. Spatial characteristics of electric current and field in large direct-current corona, *Proc. R. Soc. Lond.* 1993; **441**:125–146.

Alston LL. *High Voltage Technology.* Oxford, UK: Oxford University Press, 1968.

Amoruso V, Lattarulo F. Investigation on the Deutsch assumption: Experiment and theory. *IEE Proc. Sci. Meas. Technol.* 1996; **143**(5):334–339.

Amoruso V, Lattarulo F. A graphical approach to the unipolar point-to-plane corona. *J. Electrostat.* 1997; **39**:41–51.

Amoruso V, Lattarulo F. Ion flow from two-conductor bundle emitters. *J. Electrostat.* 1998; **43**:1–18.

Amoruso V, Lattarulo F. An improved method to explain some Warburg law deficiencies. *J. Electrostat.* 2001; **51–52**:307–312.

Amoruso V, Lattarulo F. Deutsch hypothesis revisited. *J. Electrostat.* 2005; **63** (6-10):717–721.

Atten P. Method General de Resolution du Probleme du Champ Electrique Modifiée par une Charge d'Espace Unipolaire Injectée. *Rev. Gen. Electr.* 1974; **83**(3), 143–153.

Beattie J. The Positive Glow Discharge. PhD thesis. Waterloo, Ontario, Canada: University of Waterloo, 1975.

Bennet A. *Lagrangian Fluid Dynamics.* Cambridge, UK: Cambridge University Press, 2006.

Beuthe TG, Chang JS. Gas discharge phenomena. In: Chang JS, Kelly AJ, Crowley JM, editors. *Handbook of Electrostatic Processes.* New York: Marcel Dekker, 1995, pp. 147–193.

Bhalla MS, Craggs JD. Measurement of ionization and attachment coefficients in sulphur hexafluoride in uniform fields. *Proc. Phys. Soc. London* 1962; **80**: 151–160.

Bian XB, Janfeng Hui, Youg Chen, Liming Wang, Zhicheng Guan, MacAlpine M. Simulation of Trichel streamer pulse characteristics at various air pressures and humidity. In: *Proceedings of the IEEE Conference on Electrical Insulation and Dielectric Phenomena,* October 18–21, 2009, Virginia Beach, VA, pp. 572–575.

Bouziane A, Hidaka K, Jones JE, Rowlands AR, Tapiamacioglu MC, Waters RT. Paraxial corona discharge. II. Simulation and analysis. *IEE Proc. - Sci. Measurement Technol.* 1994; **141**(3):205–214.

Boyd HA, Bruce FM, Tedford DJ. Sparkover in long uniform field gaps. *Nature (London)* 1966; **210**(5037):719–720.

Bruce FM. Calibration of uniform field spark gaps for high-voltage measurements at power-frequencies. *J. Inst. Electr. Eng.* 1953; **100**:145–153.

Canadas G, Canadas P, Dupuy J, Genet J, Marsan J, *Proceedings of the 12th International Conference on Phenomena in Ionized Gases, Eindhoven, 1975*. Amsterdam: North Holland, 1975, p. 202.

Chen M, Rong-de L, Bang-jao Y. Surface aerodynamic model of the lifter. *J. Electrostat.* 2013; **71**(2):134–139.

Ciric IR, Kuffel E. On the boundary conditions for unipolar DC corona field calculation. In: *Fourth International Symposium on High Voltage Engineering*, September 5–9, 1983, Athens, paper 13–07.

Ciric IR, Kuffel E. New analytical expressions for calculating unipolar DC corona losses. *IEEE Trans. PAS* 1982; **PAS-101**(8):2988–2994.

Cross J, Beattie J. Positive glow corona in quasi-uniform fields. *Can. Electr. Eng. J.* 1980; **5**(3):22–31.

Crowe RW, Devins JC. On the electric breakdown of electronegative gases. *Annual Report on NRC Conference Electrical Insulation*. National Research Council, Publication 396, 1956, Washington, DC, pp. 1–4.

D'Amore M, Daniele V, Ghione G. New simplified expressions for ionized fields and corona losses of HVDC monopolar transmission lines. *L'Energia Elettrica* 1986; **3**:97–104.

Davies M, Goldman A, Goldman M, Jones JE. Developments in the theory of corona corrosion for negative point-plane discharges in air. In: *Proceedings of the 18th International Conference on Phenomena in Ionized Gases,* July 13–17, 1987, Swansea, UK, p. 656.

Davis JL, Horburg JF. HVDC transmission line computations using finite element and characteristic methods. *J. Electrostat.* 1986; **16**:96–102.

Deutsch W. Uber die Dichteverteilung unipolarer Ionenstrome. *Ann. Phys.* 1933; **16**(5):588–612.

Dulikravich GS, Lynn SR., Unified electro-magneto-fluid dynamics (EMFD): Introductory concepts. *Int. J. Non-Linear Mech.* 1997a; **32**(5):913–922.

Dulikravich GS, Lynn SR., Unified electro-magneto-fluid dynamics (EMFD): A survey of mathematical models. *Int. J. Non-Linear Mech.* 1997b; **32**(5): 923–932.

Durand E. *Electrostatique*, Vol. 1. Paris: Masson Ed., 1964.

Eiceman GA, Karpas Z. *Ion Mobility Spectrometry*, 2nd ed. Boca Raton, FL: CRC Taylor and Francis, 2005.

EPRI Report, EL-2257. *DC Conductor Development*. Palo Alto, CA, 1982.

Felici N. Recent advances in the analysis of D. C. ionised electric fields—Part I. *Dir. Curr.* 1963; **8**:252–278.

Felici NJ. Electrostatics and hydrodynamics. *J. Electrostat.* 1978; **4**(2):119–129.

Gary B, Johnson JB. Degree of corona saturation for HVDC transmission lines. *IEEE Trans. Power Delivery* 1990; **5**(2):695–707.

Geballe R, Reeves ML. A condition on uniform field breakdown in electron-attaching gases. *Phys. Rev.* 1953; **92**(4):867–868.

Goldman A, Selim EO, Waters RT. In: *Proceedings of the 5th GD, IEE Conference Publication,* London 1978, No. 189, Vol. 1, pp. 146–149.

Goldman A, Goldman M, Jones JE, Yumoto M. Current distributions on the plane for point-to- plane negative coronas in air, nitrogen and oxygen. In: *Proceedings of the IX Int. Conference on Gas Discharges and Their Applications,* September 19–23, 1988, Venezia, Italy, pp. 197–200.

Hammond P. *Energy Methods in Electromagnetism.* Oxford, UK: Clarendon Press, 1981.

Hara M, Hayashi N, Shiotsuki K, Akazaki M. Influence of wind and conductor potential on distributions of electric field and ion current density at ground level in DC high voltage line to plane geometry. *IEEE Trans. PAS* 1982; **PAS-101**(4):803–814.

Harrison MA, Geballe R. Simultaneous measurement of ionization and attachment coefficients. *Phys. Rev.* 1953; **91**:1–7.

Hartmann G. Theoretical evaluation of Peek's law. *IEEE Trans. Ind. Appl.* 1984: **IA-20**(6):1647–1651.

Helmholtz HLF. Über Integrale der hydrodynamischen Gleichungen, welche den Wirbelbewegungen entsprechen. *J. Reine Angew. Math.* 2009; **1858**(55):25–55.

Henson BL. A derivation of Warburg's law for point-to-plane coronas. *J. Appl. Phys.* 1981; **52**:3921–3923.

Henson BL Derivation of the current–potential equation for steady point-to-plane corona. *J. Appl. Phys.* 1982; **53**:3305–3307.

Hermstein W. Die Stromfaden-Entladung und ihr Ubergang in das Glimmen. *Arch. Elektrotech.* 1960; **45**:209–224.

Henson BL. A space-charge region model for microscopic steady coronas from points. *J. Appl. Phys.* 1981; **52**:709–715.

Heylen ED. A comparison between sparking formulae for a uniform electric field in air. *J. Phys. D: Appl. Phys.* 1973; **6**(4):L25–L26.

Horenstein MN. Computation of corona space charge, electric field, and V–I characteristics using equipotential charge shells. *IEEE Trans. Industry Appl.* 1984; **IA-20**(6):1607–1620.

Horenstein MN. Measurement of electrostatic fields, voltages, and charges. In: Chang JS, Kelly AJ, Crowley JM, editors. *Handbook of Electrostatic Processes.* New York: Marcel Dekker, 1995, pp. 225–246.

Holzer W. Über den Stoßdurchschlag der Luft im gleichförmigen Felde bei größeren Elektrodenabständen. *Archi Elektrotech.* 1932; **26**:865–874.

IEEE Std. 539—2005. *IEEE Standard Definitions of Terms Relating to Corona and Field Effects of Overhead Power Lines,* pp. 1–41.

Ieta A, Kucerovsky Z, Greason WD. Laplacian approximation of Warburg distribution. *J. Electrostat.* 2005; **63**:143–154.

Intra P, Tippayawong N. Progress in unipolar corona discharger designs for airborne particle charging: A literature review. *J. Electrostat.* 2009; **67**(4):605–615.

Janischewskyj W, Sarma Maruvada P, Gela G. Corona losses and ionized fields of HVDC transmission lines. *CIGRE 1982*, paper 36-09-O, pp. 1–10.

Johnson GB, Zaffanella LE. Techniques for measurements of the electrical environment created by HVDC transmission lines. In: *Proceedings of the Fourth International Symposium on High Voltage Engineering,* September 5–9, 1983, Athens, Greece, Paper 13.05.

Jones JE. On the drift of gaseous ions. *J. Electrostat.* 1992; **27**:283–318.

Jones JE. On the global variational principles for corona discharges with particular reference to the active glow region. *J. Phys. D: Appl. Phys.* 2000; **33**:389–395.

Jones JE. On the interaction of adjacent coronae: A description of the electric field behaviour on the planar line beneath twin coronating negative DC points in air. *IEE Proc. Sci. Meas. Technol.* 2006; **153**(2):81–92.

Jones JE, Davies M. A critique of the Deutsch assumption. *J. Phys. D: Appl. Phys.* 1992; **25**:1749–1759.

Jones JE, Davies M, Goldman A, Goldman M. A simple analytic alternative to Warburg's law. *J. Phys. D: Appl. Phys.* 1990; **23**:542–552.

Kaune WT, Gillis MF, Weigel RJ. Technique for estimating space-charge densities in systems containing air ions. *J. Appl. Phys.* 1983; **54**(11):6267–6273.

Khalifa M, Abdel-Salam M. Calculating the surface fields of conductors in corona. *Proc. IEE* 1973; **120**(12):1574–1575.

Kondo Y., Miyoshi Y. Pulseless corona in negative point to plane gap. *Jpn. J. Appl. Phys.* 1978; **17**(4):643–649.

Landau LD, Lifshitz EM. *The Classical Theory of Fields,* revised 3rd ed. Reading, MA: Addison-Wesley, 1971.

Landau LD, Lifschitz EM, Pitaevskii LP. *Electrodynamics of Continuous Media.* Oxford, UK: Elsevier, 2008.

Lattarulo F, Amoruso V. A combined electrostatic–electrodynamic approach to lightning pre-stroke phenomena and related EMC problems. In: Lattarulo F, editor. *Electromagnetic Compatibility in Power Systems.* Oxford, UK: Elsevier, 2007, pp. 1–41.

Lattarulo F, Di Lecce F. Experimental analysis of DC corona on unipolar transmission lines. *IEE Proc.* 1990; **137C**(1):53–62.

Lemire S, Vo HD, Benner MW. Performance improvement of axial compressors and fans with plasma actuation. *Int. J. Rotating Machinery.* 2009; ID 247613.

Li JR, Wintle HJ. Unipolar corona space charge in wire-plane geometry: A first principles numerical computation. *J. Appl. Phys.* 1989; **65**(12):4617–4624.

Li W, Zhang B, Zeng R, He J. Discussion on the Deutsch assumption in the calculation of ion-flow field under HVDC bipolar transmission lines. *IEEE Trans. Power Delivery* 2010; **25**(4):2759–2767.

Loeb LB. *Fundamental Processes of Electrical Discharge in Gases*. New York: Wiley, 1939.

Loeb LB. *Electrical Coronas—Their Basic Physical Mechanisms*. Berkeley, CA: California University Press, 1965.

Loeb LB, Meek JM. *The Mechanism of the Electrical Spark*. Stanford, CA: Stanford University Press, 1941.

Malik NH. Streamer breakdown criterion for compressed gases. *IEEE Trans. Electr. Insul.* 1981; **16**(5):463–467.

Marchi E, Rubatta A. *Meccanica dei Fluidi. Principi ed Applicazioni*. Torino, Italy: UTET, 1981.

Maruvada S. P. *Corona performance of high-voltage transmission lines, RSP*, John Wiley and Sons Ltd., England, 2000.

Mason EA, McDaniel EW. *Transport Properties of Ions in Gases*. New York: John Wiley & Sons, 1988.

McKnight R. H., Kotter F. R., Misakian M. Measurement of ion current density at ground level in the vicinity of high voltage DC transmission lines. *IEEE Trans. Power Apparatus and Systems* 1983; **PAS-102**:934–941.

McLean KL, Ansari IA. Calculation of the rod-plane voltage/current characteristics using the saturated current density equation and warburg's law. *IEE Proc. Part A* 1987, **134**(10):784–788.

Meek JM, Craggs JD. *Electrical Breakdown of Gases*. Oxford, UK: Oxford University Press, 1953.

Metwally IA. Factors affecting corona on twin-point gaps under DC and AC HV. *IEEE Trans. Dielectr. Electr. Insul.* 1996; **3**(4):544–553.

Misakian M. Generation and measurement of DC electric fields with space charge. *J. Appl. Phys.* 1981; **52**(5):3135–3144.

Misakian M, McKnight RH. Calibration of aspirator-type ion counters and measurement of unipolar charge densities. *J. Appl. Phys.* 1987; **61**(4):1276–1287.

Misakian M, Anderson WE, Laug OB. Drift tubes for characterizing atmospheric ion mobility spectra using AC, AC-pulse, and pulse time-of-flight measurement techniques. *Rev. Sci. Instrum.* 1989; **60**(4):720–729.

Moffatt HK. *Magnetic Field Generation in Electrically Conducting Fluids*. London: Cambridge University Press, 1978.

Moon P, Spencer DE. *Field Theory for Engineers*. Princeton, NJ: Van Nostrand, 1961.

Morse PM, Feshbach H. *Methods of Theoretical Physics, Part I*. New York: McGraw-Hill, 1953.

Nouri H, Zouzou N, Moreau E, Dascalescu L, Zebbudj Y. Effect of relative humidity on current-voltage characteristics of an electrostatic precipitator. *J. Electrostat.* 2012; **70**(1):20–24.

Ohashi S, Hidaka K. A method for computing current density and electric field in electrical discharge space charge using current flow-line coordinate. *J. Electrostat.* 1998; **43**(2):101–114.

Panton RL. *Incompressible Flow*. New York: John Wiley & Sons, 1984.

Pappas PT. Electrodynamics from Ampere to now: Furure perspectives. In: Lathtakia A editor. *Essay on the Formal Aspects of Electromagnetic Theory*. Singapore: World Scientific, 1993, pp. 287–309.

Pedersen A. On the electrical breakdown of gaseous dielectrics. An engineering approach. *IEEE Trans. Electr. Insul.* 1989; **24**(5):721–739.

Penfield P, Jr, Hauss HA. *Electrodynamics of Moving Media*. Research monograph, no. 40. Cambridge, MA: The MIT Press, 1967.

Popkov VI. On the theory of unipolar DC corona. *Elektrichestvo* 1949; **1**:33–48.

Popkov VI. Theory of corona discharge in gas for constant voltage potential. *Izv. Akad. Nauk USSR* 1953; **5**:664.

Popkov VI, Ryabaya SI. Distribution of current of unipolar corona at the non-corona and corona electrodes. *Elektrichestvo* 1974; **11**:45.

Popkov VI, Bodganova NB, Pevchev BG. Electric field intensity on the surface of a positive electrode under conditions of an oncoming stream of negative ions. *Izv. Akad. Nauk USSR* 1978; **16**:96.

Qin BL, Sheng JN, Yan Z. Accurate calculation of ion flow field under HVDC bipolar transmission lines. *IEEE Trans. Power Delivery* 1988; **3**(1):368–376.

Raether HZ. The electron avalanche and its development. *Appl. Sci. Res. Sect. A* 1956; **5**(1):23–33.

Raether H. *Electron Avalanches and Breakdown in Gases*. Washington, DC: Butterworth Press, 1964.

Revercomb HE, Mason EA. Theory of plasma chromatography/gaseous electrophoresis—A review. *Anal. Chem.* 1975; **47**(7):970–983.

Ritz H. Durchschlagfeldstaerke des homogenen Feldes in Luft. *Arch. Elektrotech.* 1932; **26**:219–232.

Saha MH. Ionization in the solar chromosphere. *Philos. Mag.* 1920; **40**:472–488.

Sarma Maruvada P. *Corona Performance of High-Voltage Transmission Lines*. Baldock, UK. RSP Studies Press, 2000.

Sarma Maruvada P. Electric field and ion current environment of HVDC transmission lines. *IEEE Trans. Power Delivery* 2012; **27**(1):401–410.

Sarma Maruvada P, Janischewskyj W. Analysis of corona losses on DC transmission lines: I-unipolar lines. *IEEE Trans. Power Appar. Syst.* 1969; **88**:718–731.

Selim EMO, Waters RT. Static probe for electrostatic field measurement in the presence of space charge. *IEEE Trans.* 1980; **IA-16**(3):458–463.

Shin WT, Sung NC. Organic pollutants degradation using pulseless corona discharge: Application in ultrapure water production. *Environ. Eng. Res.* 2005; **10**(3):144–154.

Sigmond RS. Corona discharge. In: Meek JM, Craggs JD, editors. *Electrical Breakdown of Gases*. New York: J. Wiley & Sons, 1978, pp. 319–384.

Sigmond RS. Simple approximate treatment of unipolar space-charge-dominated coronas: The Warburg law and the saturation current. *J. Appl. Phys.* 1982; **53**: 891–898.

Sigmond RS. The unipolar corona space charge flow problems. *J. Electrostat.* 1986; **18**(3):249–272.

Solymar L, Walsh D. *Lectures of the Electrical Properties of Materials.* Oxford, UK: Oxford University Press, 1993.

Stratton JA. *Electromagnetic Theory.* Hoboken, NJ: Wiley-IEEE Press, 2007.

Takuma T, Ikeda T, Kowamoto T. Calculation of ion flow fields of HVDC transmission lines by the finite element method. *IEEE Trans. PAS* 1981; **PAS-100**(12):4802–4810.

Townsend JS. *The theory of ionization of gases by collision.* London: Constable and Co., 1910.

Townsend JSE. *Electricity in Gases.* Oxford, UK: Clarendon Press, 1914.

Trichel GW. The mechanism of the negative point to plane corona near onset. *Phys. Rev.* 1938; **54**(12):1078–1084.

Trinh NG, Jordan JB. Modes of corona discharges in air. *IEEE Trans. PAS* 1968; **PAS-87**(5):1207–1215.

Tsyrlin LE. Condition for the preservation of the geometrical pattern of the electrostatic field on the appearance of space-charge (in Russian). *Sov. Phys. -Tech. Phys.* 1957; **30**:2439.

Tsyrlin LE. *Sov. Phys. -Tech. Phys.* 1958; **31**:1470.

Usynin GT. The calculation of the field and characteristics of unipolar DC corona discharge (wire parallel to plane) (in Russian). *Izv. Akad. Nauk-SSSR* 1966; **4**:56–70.

Walden PZ. Über organische Lösungs-und Ionisierungsmittel. III. Teil: Innere Reibung und deren Zusammenhang mit dem Leitvermögen. *Z. Phys. Chem.* 1906; **55**:207–246.

Walsh PJ, Pietrowskj KW, Sigmond RS. The negative corona current distribution for a long pin-to-plane geometry. *Photographic Sci. Eng.* 1984; **28**(3):101–108.

Wannier GH. Motion of gaseous ions in strong electric fields. *Bell System Tech. J.* 1953; **32**(1):170–254.

Warsi ZUA. *Fluid Dynamics. Theoretical and Computational Approaches.* Boca Raton, FL: CRC Press, 1999.

Waters RT, Rickard TES, Stark WB. The structure of the impulse corona in a rod/plane gap: I. The positive corona. *Proc. R. Soc.* 1970; **A315**:1–25.

Waters RT, Rickard TES, Stark WB. *Electric Field Measurements in DC Corona Discharges.* 1972 (IEE Conf. Publ. 90), pp. 188–190.

Yang Y, Lu J, Lei Y. A calculation method for the electric field under double-circuit HVDC transmission lines. *IEEE Trans. Power Delivery* 2008; **23**(4):1736–1742.

Zheng Y, He J, Zhang B, Li W, Zeng R. Ion flow effects on negative direct current corona in air 2010. *Plasma Chem. Plasma Process* **30**:55–73.

INDEX

Filamentary Ion Flow: Theory and Experiments, First Edition. Edited by Francesco Lattarulo and
Vitantonio Amoruso.
© 2014 by The Institute of Electrical and Electronics Engineers, Inc. Published by 2014 John Wiley & Sons, Inc.